CARS AND CLIMATE CHANGE

INTERNATIONAL ENERGY AGENCY

2, RUE ANDRÉ-PASCAL, 75775 PARIS CEDEX 16, FRANCE

The International Energy Agency (IEA) is an autonomous body which was established in November 1974 within the framework of the Organisation for Economic Co-operation and Development (OECD) to implement an international energy programme.

It carries out a comprehensive programme of energy co-operation among twenty-three* of the OECD's twenty-four Member countries. The basic aims of the IEA are:

i) co-operation among IEA participating countries to reduce excessive dependence on oil through energy conservation, development of alternative energy sources and energy research and development;

ii) an information system on the international oil market as well as consultation with oil companies;

iii) co-operation with oil producing and other oil consuming countries with a view to developing a stable international energy trade as well as the rational management and use of world energy resources in the interest of all countries;

iv) a plan to prepare participating countries against the risk of a major disruption of oil supplies and to share available oil in the event of an emergency.

IEA participating countries are: Australia, Austria, Belgium, Canada, Denmark, Finland, France, Germany, Greece, Ireland, Italy, Japan, Luxembourg, the Netherlands, New Zealand, Norway, Portugal, Spain, Sweden, Switzerland, Turkey, the United Kingdom, the United States. The Commission of the European Communities takes part in the work of the IEA.

ORGANISATION FOR ECONOMIC CO-OPERATION AND DEVELOPMENT

Pursuant to Article 1 of the Convention signed in Paris on 14th December 1960, and which came into force on 30th September 1961, the Organisation for Economic Co-operation and Development (OECD) shall promote policies designed:

— to achieve the highest sustainable economic growth and employment and a rising standard of living in Member countries, while maintaining financial stability, and thus to contribute to the development of the world economy;

— to contribute to sound economic expansion in Member as well as non-member countries in the process of economic development; and

— to contribute to the expansion of world trade on a multilateral, non-discriminatory basis in accordance with international obligations.

The original Member countries of the OECD are Austria, Belgium, Canada, Denmark, France, Germany, Greece, Iceland, Ireland, Italy, Luxembourg, the Netherlands, Norway, Portugal, Spain, Sweden, Switzerland, Turkey, the United Kingdom and the United States. The following countries became Members subsequently through accession at the dates indicated hereafter: Japan (28th April 1964), Finland (28th January 1969), Australia (7th June 1971) and New Zealand (29th May 1973). The Commission of the European Communities takes part in the work of the OECD (Article 13 of the OECD Convention).

TABLE OF CONTENTS

TABLES

FIGURES

FOREWORD

When the IEA was established in 1974, the transport sector in the OECD accounted for a quarter of final energy use — less than either industry or the residential/commercial sector. In 1990 transport was responsible for almost a third of final energy use, and in some countries it was the largest final energy use sector. Oil makes up 99% of the energy use in transport. Since the late 1970s IEA governments, concerned about energy security, have sought means of promoting energy efficiency and fuel flexibility in this sector.

In recent years concern about the environmental effects of transport has been increasing. Some of the environmental consequences of transport are obvious to most people in OECD Member countries. Cars, unlike power stations and refineries, produce air pollution in places where people live and work. Policies to reduce the problems caused by motorisation have so far focused on the social effects: congestion, noise, accidents and air pollution.

The recent rise in concern about anthropogenic global warming has added to the number of issues policy makers have to balance. Policy related to cars is especially difficult because of their social and economic importance. The problem for governments, then, is to find ways of managing the social and environmental problems cars cause while preserving the pleasure, freedom and mobility they afford.

IEA Member countries are adopting a wide range of strategies to address these issues. They include new car technology development, investment in public transport infrastructure, creative developments in urban planning and more direct measures to manage growth in traffic, energy use and pollution.

At a meeting in June 1991, energy ministers from IEA countries noted with concern the continued growth of oil demand in the transport sector, and urged the acceleration of efficiency gains in this sector. They undertook to work with transport ministers and with their national industries to increase the availability of fuel-efficient vehicles, stimulate greater use of mass transit, encourage switching from road to rail for goods transport and remove barriers to the use of alternative fuels.

The IEA Secretariat does not aspire to offer solutions. Governments will choose their own paths, balancing technological solutions with economic and social measures. This report confines itself to demonstrating methodologies whereby measures can be systematically assessed and compared.

It addresses itself to a single problem — greenhouse gas emissions — and a relatively small number of possible solutions: improved energy efficiency and alternative transport fuels. The life-cycle analysis methodology used here could be applied to other environmental problems and other solutions. This analysis has been based on a very small number of scenarios, and Member countries are urged to carry out their own analyses to reflect their preferred scenarios.

The IEA Secretariat prepared this study under guidance from a group of experts drawn from government and industry. *Cars and Climate Change* is published under my authority as Executive Director of the IEA and does not necessarily reflect the views or policies of the IEA or its Member countries' governments.

Helga Steeg
Executive Director

EXECUTIVE SUMMARY

The transport sector is an essential element in the process of creating and consuming wealth. Popularisation of the motor car in particular has been important in the process of industrialisation and economic growth. At the same time there is an emerging consensus among OECD governments that policies are required to address some of the adverse social and environmental effects of motor vehicles. Traffic can be detrimental to quality of life, especially in cities, through the risks, noise and air pollution it causes. Vehicles are also a major source of greenhouse gas emissions. In the context of continuing growth in car use, stabilisation of the emissions poses a major challenge.

Cars and Climate Change contributes to the analysis of the technical potential, economic potential and market potential for emission reduction in the transport sector through increased efficiency and fuel substitution. The car market and its related fuel supply and infrastructure systems are examined, along with policies that might effect beneficial changes in the market.

ENERGY USE AND EMISSIONS

Energy use in OECD transport nearly tripled between 1960 and 1990. The growth rate for emissions of carbon dioxide (CO_2) was virtually the same, though other transport emissions have been decreasing. Transport, including international marine bunker fuel use, is now responsible for more than one-third of OECD final energy use. It is the largest final energy use sector and the share is growing. It is also the sector that has been the least responsive to policy makers' attempts to encourage energy efficiency and fuel flexibility.

Over the last 20 years, all transport modes except seagoing ships carried increasing levels of passenger traffic. The increase in rail and bus travel has been slight, and most of the additional land-based travel is by car. Of the passenger transport modes, air travel, which has the highest energy use and greenhouse gas emissions per passenger-kilometre, has increased fastest. Its growth rate is matched by that of road freight traffic. Air travel and road freight, which are causing increased concern in terms of both energy use and the environment, will be the focus of future IEA studies.

Of the land-based passenger transport modes, car travel is the most energy intensive. At typical seat occupancy levels, buses and trains use less energy per passenger-kilometre. Gasoline-powered cars, in aggregate, consume more energy than any other type of vehicle, and produce more greenhouse gas emissions.

REDUCING GREENHOUSE GAS EMISSIONS FROM CARS

For this study the IEA has used a life-cycle emission model that takes into account upstream emissions in considerable detail. Fuel supply is analysed, including raw material extraction, transport, processing and fuel distribution. Similarly, the model calculates emissions in vehicle production, from raw material extraction, transport and processing to vehicle manufacture. The model can be used to examine the effects on emissions of vehicle and engine design, of switching to alternative fuels and of using electric vehicles. The model also takes account of emissions other than CO_2, weighting them according to their greenhouse forcing[1] and how long they stay in the atmosphere.

About 72% of greenhouse gases from cars are emitted from the tailpipe during vehicle operation; 17-18% of car life-cycle emissions arise from fuel extraction, processing and distribution; a further 10% come from vehicle manufacture[2]. For cars with below-average annual kilometrage, the emissions in vehicle manufacture become more significant as a proportion of life-cycle emissions. The reverse holds for cars with above-average kilometrage.

Exhaust emission control devices are expected to be installed on most cars throughout the OECD by about 2005. Catalytic converters reduce emissions of carbon monoxide, volatile organic compounds (VOCs) and nitrogen oxides (NO_x). However, they increase emissions of CO_2 and nitrous oxide (N_2O).

Greenhouse gas emissions can be reduced through:

- **Energy efficiency improvements.** Lower fuel use — for example, as a result of improved aerodynamic design — can reduce emissions throughout the fuel and vehicle life-cycle.

- **Fuel switching.** Alternative energy carriers can result in lower life-cycle CO_2 emissions because they contain less carbon, or because they contain carbon absorbed by plants from the atmosphere. Some alternative fuels can give higher engine efficiency than gasoline. Life-cycle analysis is particularly important in examining the potential benefits of alternative fuels.

These two measures can complement each other. Improvements in gasoline vehicle design are clearly applicable to most alternative-fuel vehicles. Similarly, the vehicle design improvements that will be necessary to develop a viable electric vehicle can be used in gasoline vehicle production.

---

1. Effect on global radiative balance per unit mass.
2. Emissions of chlorofluorocarbons (CFCs) and emissions associated with vehicle disposal vary widely between countries and are not treated in this report.

ENERGY EFFICIENCY IMPROVEMENTS

Technical Potential. Technology is available that would improve car fuel economy by a factor of three or more. This could not be done without reducing performance or raising costs, however. Few of the resulting cars would be competitive in today's market.

Economic Potential. Analysis of the energy efficiency distribution of the current fleet can be used to indicate the economic potential for energy efficiency improvements: the fuel economy that would be achieved if car purchasers were to choose the model that satisfies their needs at the least overall cost. Studies in the United States and the United Kingdom indicate that the economic potential is probably at least 20% better than the current average fuel economy.

Market Potential. Many analysts have attempted to identify the market potential for fuel economy improvements — that is, the improvement that the market will produce without additional intervention. This can be done by:

- making techno-economic assessments of changes that do not affect vehicle size, performance or comfort level;
- mapping the energy efficiency distribution of cars currently being purchased and using the top 10% or 20% to indicate the potential for the fleet as a whole over the next ten to 20 years;
- using macroeconomic models to generate scenarios of the future that include energy efficiency indicators as an output.

All these approaches suggest that fuel economy may improve by 10-20% between now and 2005.

ALTERNATIVE FUELS

Some alternative fuels — diesel, LPG[1] and CNG[2], for example — can be produced with less processing than gasoline from crude oil. Synthetic fuels such as alcohols generally require more energy and more capital-intensive plant for processing. Switching fuels generally results in lower tailpipe emissions of CO_2 and other pollutants but may result in higher emissions from fuel supply. Where alternative liquid fuels are produced from gas or coal, life-cycle greenhouse gas emissions can exceed those due to gasoline use. Fuels from biomass or other renewable sources can in principle have zero life-cycle emissions. Manufacturing of vehicles using gaseous fuels that require heavy cylinders, or electric vehicles with heavy batteries, involves more energy use and emissions than that of more conventional cars.

Technical Potential. Figure 1 shows an example of the calculation of life-cycle emissions for a variety of alternative fuel options for use in North America. The options can be divided into four main groups:

- Fuels which offer little or no greenhouse gas abatement but may be attractive from the perspective of other areas of government policy. Synthetic liquid fuels using fossil fuel

---

1. Liquefied petroleum gas.
2. Compressed natural gas.

inputs, including some biomass-derived fuels, fall into this group, as do CNG used in existing vehicles (not shown in the graph) and electric vehicles using power from some existing generation mixes;

- Alternatives available now, or expected to become available by 2005, including diesel, LPG, CNG in optimised engines and electric vehicles using power from existing generation mixes; these options can reduce greenhouse gas emissions by 10-25%;

- Synthetic fuels from wood or other low-input biomass feedstocks, which are not yet technically demonstrated but could offer 60-80% greenhouse gas abatement;

- Fuels derived from completely renewable sources, including hydrogen produced by electrolysis of water using electricity generated by renewable sources; synthetic fuels from zero-input biomass feedstocks; and electric vehicles powered by electricity from renewable sources. All would mean large-scale replacement of the existing fossil-based energy system. They can result in over 80% greenhouse gas abatement.

Figure 1
Life-cycle Greenhouse Gas Emissions from Alternative-Fuel Cars, North America, 2000
(g/km of CO_2 equivalent)

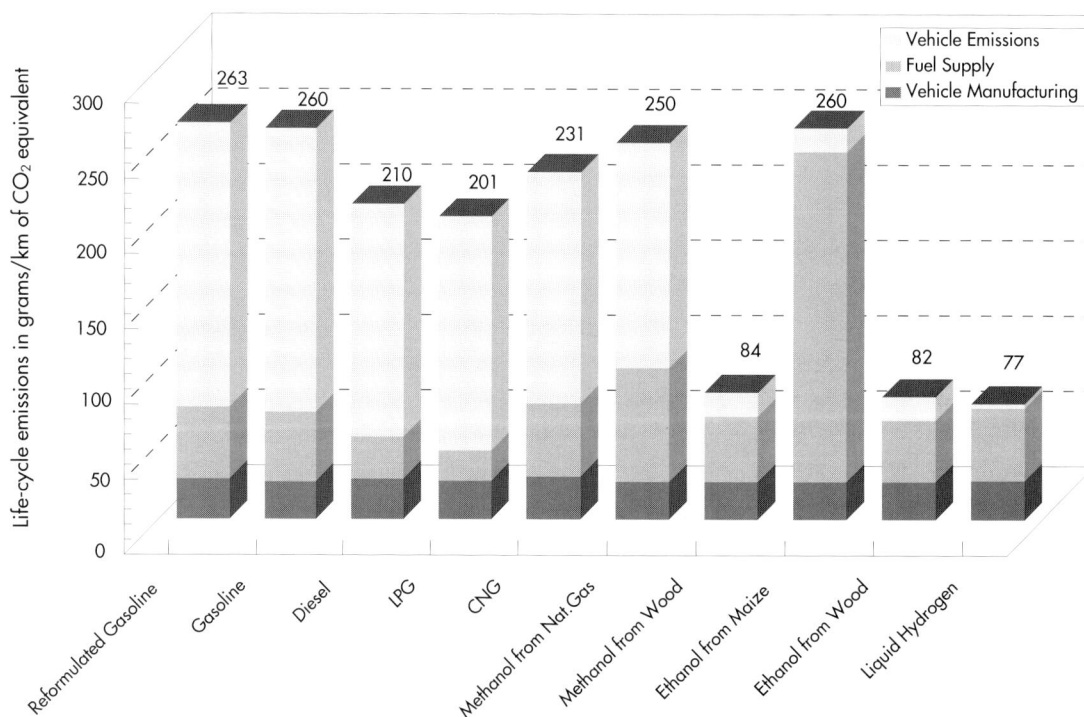

One striking result of the analysis of alternative fuels and electric vehicles is the considerable range of emission levels that could be associated with each option (see **Figure 2**). The results depend on the fuel inputs and emission levels associated with power generation and fuel conversion. Any ranking of the options will vary by region and according to the assumptions made about technology that is not yet fully developed. Even currently available options, including CNG and ethanol from maize, have considerable ranges of emissions and may result in higher life-cycle emissions than gasoline.

Figure 2
Greenhouse Gas Emission Ranges from Alternative-Fuel Cars
(Reformulated Gasoline = 100)

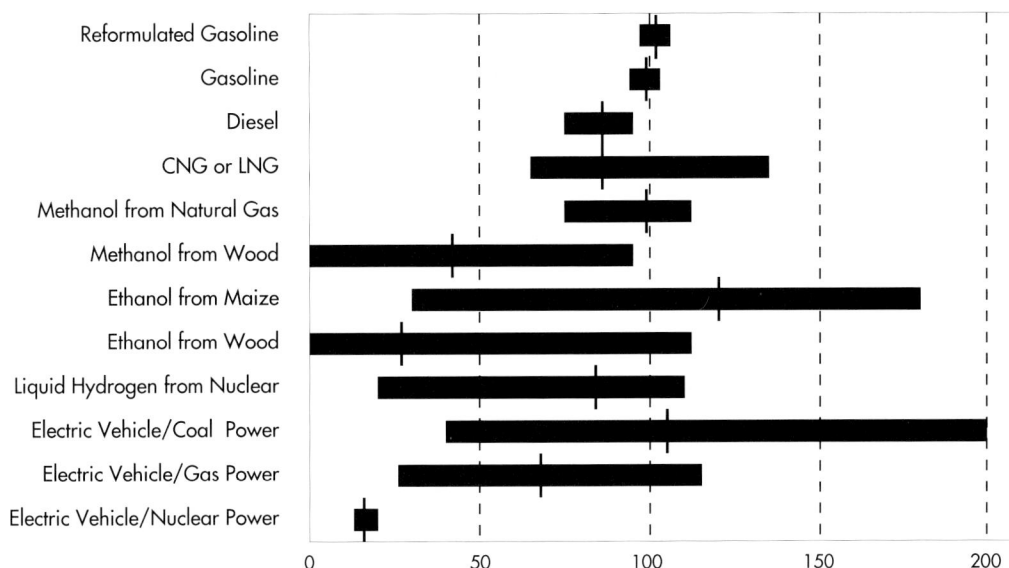

Note: This graph indicates the possible range of greenhouse gas emissions per kilometre for each alternative fuel and for electric vehicles with various power sources. The ranges can result both from uncertainties about the car technologies and from local variations between energy supply technologies and systems.

Economic Potential. The car buyer considering an alternative-fuel vehicle has to consider the cost of the vehicle, its probable operating costs and its expected resale value. In the case of fuels such as CNG or diesel, the vehicle cost is likely to be higher than that of a gasoline vehicle and the fuel costs are likely to be lower. The buyer has to make a trade-off, depending on the cost of capital, expected annual costs and kilometrage and the probable time before the car will be resold.

An earlier IEA study examined the costs and technical feasibility of using several alternative fuels (IEA, 1990b). **Figure 3** shows the cost-effectiveness of using alternative fuels to reduce greenhouse gas emissions, considering only the costs involved in fuel supply.

The current study provides a deeper economic analysis of fuels that may have significant market potential by the end of the 1990s. Costs are calculated for gasoline, diesel and CNG cars in the United States and France in 2000. **Figure 4** shows the estimated ranges of costs in each country of switching from gasoline to diesel and CNG cars at 1992 fuel prices and taxes. The fuel duties in each country have important effects on the economics of fuel switching. In France diesel is subject to lower tax than gasoline, and is likely to remain very attractive for most vehicle buyers. Tax exemptions introduced in the United States by the 1992 Energy Policy Act may make CNG attractive, at least for drivers who are unaffected by a shorter driving range.

Figure 3

Alternative Fuels' Cost-Effectiveness for Greenhouse Gas Abatement, 2000

(1987 US $ per metric ton of CO_2 equivalent)

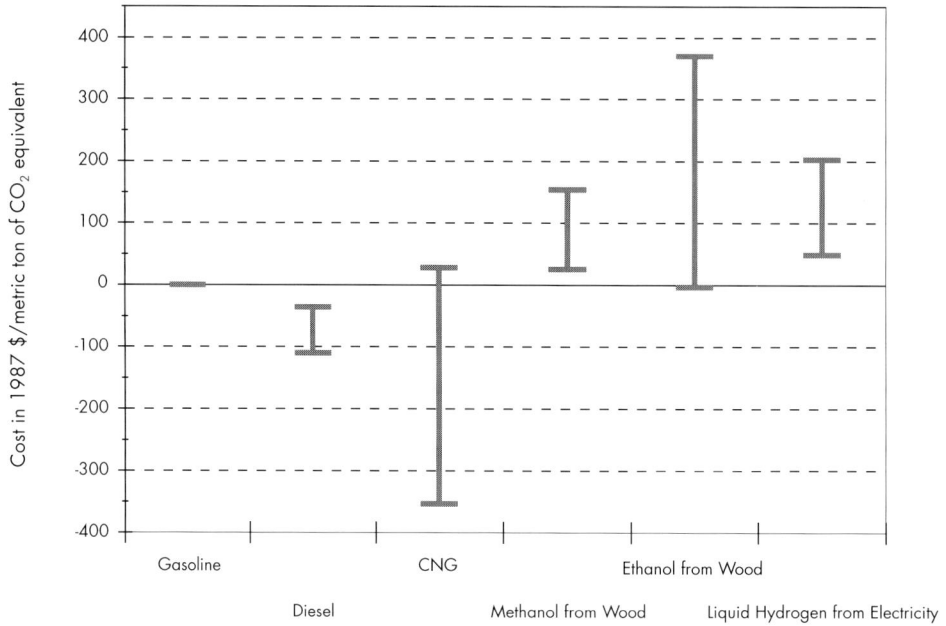

Source: IEA (1990b), Marrow and Coombs (1990), CEC (1992).

Figure 4

Cost of Switching from Gasoline to Diesel or CNG, 2000

(Levelised Cost in 1992 US cents per km)

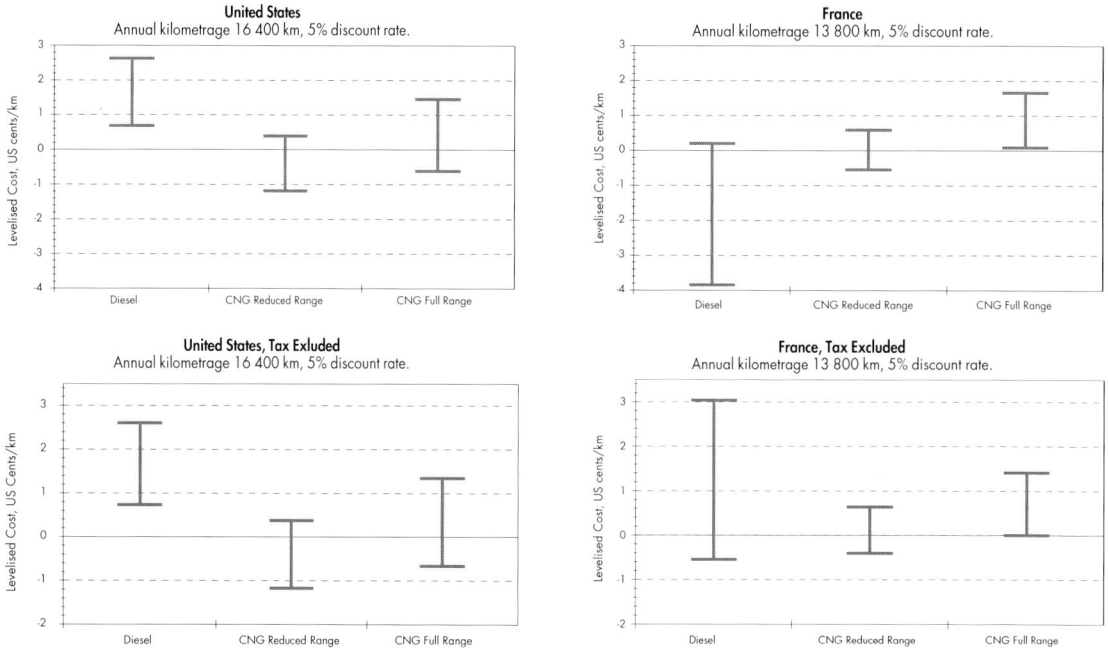

18

Market Potential. Market share projections for alternative fuels are unreliable, as there is little experience on which to base them. Macroeconomic models such as the IEA's World Energy Outlook are not designed to predict fuel switching in the long term. Econometric models with more detailed disaggregation of transport fuel demand may be more helpful in identifying possible niche markets for alternative fuels.

Market surveys have been carried out in California, where alternative fuels are being promoted by the state government. The surveys indicate that disadvantages of alternative fuels, such as uncertainty about availability, outweigh any cost advantage for most consumers. As a result the main users of alternative-fuel vehicles have tended to be fleet operators.

POLICIES FOR GREENHOUSE GAS ABATEMENT

Many OECD Member countries have adopted policies to promote alternative fuels. These policies have usually been motivated by objectives other than greenhouse gas abatement. In the United States alternative fuels are being introduced as a result of legislation that is intended mainly to reduce emissions of carbon monoxide and VOCs.

Energy-efficient vehicles are not achieving their economic potential in the car market now, and alternative-fuel vehicles appear unlikely to do so by 2005 without government intervention. This is partly due to aspects of the technologies that make them unattractive to consumers — reduced performance, uncertainty regarding fuel availability, uncertainty about the resale market. It may also be due to market imperfections, such as lack of information about new technologies or the existence of external costs and benefits associated with them.

CO_2 emissions are linked directly to fossil fuel demand. In economic terms the most efficient way to reduce emissions would be to tax all fuels, in all sectors, throughout the world, according to their carbon content. This approach, however, is unlikely to be adopted in the near future. The external cost of CO_2 emissions is not known and may be unknowable, so it is not possible to determine the tax level that would internalise the cost.

Approaches that do not depend on international agreement, such as vehicle fuel economy standards, have been widely adopted. Such standards may have the drawback of resulting in lower driving costs and hence more propensity to drive. Other indirect approaches to reducing fuel demand may have similar drawbacks. Even if they result in fuel savings, they are likely to do so at greater expense in consumer welfare than would have been incurred using carbon taxes.

A case study carried out in the Netherlands analyses the effects on traffic and emissions of several policy measures, including parking controls, fuel pricing, road pricing and public transport investment. The study also examines the effects of combinations of different types of measures. Combined measures have more effect than would be produced by adding the effects of the component measures. The use of such combinations reduces opportunities for consumers to compensate for restrictions imposed by individual measures.

POLICIES FOR SUSTAINABLE TRANSPORT

Although greenhouse gas abatement appears difficult to achieve for passenger cars, there is growing recognition of the range of problems caused by cars. Oil dependence has long been a concern of governments in OECD countries. Other issues rising in the political agenda include traffic congestion, accidents, noise and local air pollution.

These issues have relevance for greenhouse gas emissions. Policies to deal with the other problems caused by transport can also reduce greenhouse gas emissions. For example, in Europe the fitting of speed limiting devices to heavy-duty vehicles reduces not only accidents but also energy use. In California the promotion of CNG vehicles to reduce local air pollution may also result in reduced greenhouse gas emissions.

Concern about global warming adds weight to the arguments for governments to reconsider their transport policies according to the "polluter pays" principle. They should try to reduce the damage caused by transport as far as possible. Where damage cannot be reduced, transport users should be required to pay the full cost of their mobility. Yet the considerable existing government intervention affecting transport makes this task difficult.

Responsibility for acting on many problems associated with transport tends to be split between government departments. National administrations are beginning to address transport sector issues as a whole, by consultation between departments. The process is important in helping policy makers see the synergy among the different issues, and should result in more effective action to deal with each problem.

This report cannot prescribe policies or policy packages for governments. The main recommendation arising from the study is that governments should carry out and act on their own careful, comprehensive analyses of transport policy options.

CHAPTER 1

INTRODUCTION

This is a study of energy and environmental responses in the transport sector. It is the first of a number of sectoral analyses to be undertaken by the IEA as detailed follow-up to three recently completed studies[1]. The latter are a point of departure for the quantitative analysis undertaken in this study.

Transport services are an integral part of the creation and consumption of wealth in the OECD. The development and use of transport infrastructure accounts for 4-8% of GDP and 2-4% of jobs (ECMT, 1990). Rising mobility of goods and people accompanies economic growth and facilitates competition for jobs, goods and services. Consumer aspirations for car ownership and leisure travel are driving forces in the growth of the world economy.

The transport sector has an important impact on energy security. Consumption of oil in road transport almost tripled between 1960 and 1990. Transport's share in final energy use in the OECD is about 34%[2]. In addition to oil use for vehicle operation, a significant amount of energy is used in making vehicles and providing infrastructure. The energy use for car manufacture is about 15% of the energy use in the vehicle life-cycle. It represents roughly a tenth of energy use in the industry sector in the OECD.

The transport sector is responsible for over two-thirds of OECD oil consumption. Demand for oil in the sector has low price and income elasticity: consumption does not fall significantly in times of high oil prices or recession. This can have serious effects on the balance of payments in oil importing countries. In addition, the concentration of global oil supplies in a few regions raises concern regarding security of supply.

The growth in energy use and lack of fuel flexibility in transport are a source of concern. Most OECD countries are net oil importers, and the oil price has been volatile over the last 20 years.

---

1. The first (IEA, 1990a) describes the overall interaction of energy and environmental issues and suggests an analytical framework for more detailed study. The second (IEA, 1991d) provides an overview of energy-related greenhouse gas emissions, abatement options and policy instruments. The third (IEA, 1991a) analyses energy efficiency trends and opportunities in OECD Member countries and the implications for CO_2 emissions.

2. Including energy use for international marine bunkers.

Since the mid-1970s energy efficiency and fuel switching in this sector have been of interest to policy makers concerned about energy security.

Transport has also become one of the principal targets for improving environmental quality. Road transport in particular is a major contributor to urban air pollution. Land use issues, noise and congestion associated with the transport system also pose serious problems. The increase in transport demand has tended to offset any reduction in pollution from individual vehicles achieved by tightening emission standards.

While some of the problems caused by road transport appear critical, demand for cars and car use continue to rise. The trend is strong and difficult to counter, partly because the direct cost of road transport is very low, and partly because motorisation and mobility are so important in 20th century culture and economic growth. Societies and infrastructure have changed so that the private car has become almost indispensable. Policy makers find the transport sector particularly difficult to address effectively. The normal response is to build roads to cope with growth in traffic, which in turn encourages increased traffic.

In this context, greenhouse gas abatement in the transport sector would appear to be extremely difficult. Yet there may be considerable scope for it. As governments increasingly recognise the problems caused by road transport, the incentive and justification for action also increase. The addition of global warming to the list of problems may for some tip the balance in favour of policy measures that result in transport users paying directly for more of the burden they place on society.

The issues that have to be weighed in considering measures are complex. Many of the societal costs of transport are hard or impossible to quantify. The response of transport users, if these costs were to be reflected in the price of mobility, is similarly hard to predict. This report aims to provide part of the framework in which some of these issues can be addressed.

The original objectives of the study were:

- to provide a detailed inventory and life-cycle analysis of greenhouse gas emissions from the transport sector;

- to examine the cost-effectiveness of greenhouse gas abatement through efficiency improvements and alternative road transport fuels and technologies;

- to examine the cost-effectiveness of policy instruments aimed at reducing greenhouse gas emissions from the transport sector.

It was quickly found that the data needed for a detailed OECD-wide analysis was not available. Using country submissions to the IEA, it is possible to estimate overall carbon dioxide (CO_2) emissions from transport energy use, but it is not possible to obtain accurate data for all countries on emissions of other gases, or breakdowns of fuel use by vehicle types and applications. This kind of data is available from a very few countries where detailed transport energy analyses have been carried out. Similarly, detailed information on vehicle use patterns — for example, kilometres driven per year, broken down by vehicle type, engine size and age — is available for few countries, and the data quality is variable.

A considerable amount of information has been collected on car technologies and markets and on the potential for technical change in cars. There is little information available on other vehicles, however, apart from a few studies in individual countries.

As a result of these limitations, a more tightly focused case study approach was adopted for this study. This report presents a methodology for technology policy analysis, focusing on alternative fuels for cars in North America and OECD Europe. The study is divided into three parts. **Part One** examines the contribution of the transport sector to greenhouse gas emissions and compares the different transport modes. It provides a background for the main analytical work presented in the study.

Chapter 2 briefly outlines the nature and magnitude of greenhouse gas emissions from fossil fuel use in the OECD. Using the IEA/OECD greenhouse gas emission inventory, it shows the importance of transport as a greenhouse gas source. **Chapter 3** discusses the relative contribution of the main transport modes: road, air, rail and water. It looks at the trends in traffic levels and technology development in each mode. Road and air transport are identified as the subsectors with the largest growth rates, both in traffic and in energy use. Within the road subsector, growth is particularly rapid in freight transport. Road and air transport contribute a growing share of CO_2 emissions and a disproportionate share of other pollutants.

Part Two is a quantitative analysis of greenhouse gas emissions from light-duty road vehicles and opportunities for abatement. **Chapter 4** provides background on current light-duty vehicles. It covers the nature of vehicle and engine technology and examines the technical factors that affect greenhouse gas emissions. **Chapter 5** examines probable improvements in vehicle energy efficiency over the next 15 years. It begins to consider some of the most difficult problems associated with technical change in the automotive industry. The chapter identifies the difference between:

- **technical potential,** the technically feasible level of greenhouse gas abatement;

- **economic potential,** the level of abatement given technical changes that are financially attractive;

- **market potential,** what is likely to happen given the interplay of technical change, markets and personal preferences.

A number of analyses have already been made of the potential for vehicle fuel economy improvements, and this study does not attempt to add to them. Studies of technical potential provide a view of possible developments in the market in the very long term, beyond the time frame of this study. Studies of economic potential are of more interest, as they identify technology that could feasibly be introduced in the next ten to twenty years. By contrasting the results of this type of analysis with the results of market analysis, it is possible to identify areas where intervention might be most effective.

The market potential can be estimated from car manufacturers' views of the direction in which technology and consumer tastes are likely to develop. It can also be examined using macroeconomic models, which do not attempt to disaggregate the decision making behind consumers' technology choices. The IEA's World Energy Outlook provides this type of analysis. The results of the model are compared with detailed industry studies in **Chapter 5**.

Chapter 6 analyses the technical potential for greenhouse gas abatement using alternative transport fuels and electric vehicles. A spreadsheet-based, life-cycle analysis is used to compare overall greenhouse gas emissions in the various options. The analysis is based on conditions in two OECD regions: North America and OECD Europe.

In **Chapter 7**, a comparison is made of the costs of operating vehicles on gasoline, diesel and compressed natural gas (CNG). A number of fuel and vehicle price scenarios are considered for France and the United States. The analysis reveals conditions under which the different fuels would be financially attractive to vehicle buyers. This approach can be used to give a first-order estimate of the cost of fuel switching per metric ton of emissions avoided. A full analysis of the cost of carbon abatement using alternative fuels would require a more detailed study using econometric models.

Although heavy-duty vehicles have not been analysed in detail in this report, it is recognised that they pose significant and growing environmental problems. **Chapter 8** presents a brief analysis of the technical opportunities for greenhouse gas abatement in heavy-duty vehicles.

Part Three focuses on market and policy issues. **Chapters 9 and 10** examine the issues currently influencing vehicle energy efficiency, and the alternative transport fuel markets in North America and OECD Europe. It is apparent from analysis of the markets that technology is not being introduced on the grounds of cost-effectiveness alone.

Chapter 11 looks in more detail at the way the transport market works, and areas in which the market is not functioning well. It draws on a number of case studies to give an indication of the potential of different policy measures to deal with some of the wider problems caused by transport. An analysis is given of the implications of these measures for vehicle fuel economy and the introduction of alternative fuels.

PART I

GREENHOUSE GAS EMISSIONS FROM THE TRANSPORT SECTOR

CHAPTER 2

TRANSPORT AND THE ENVIRONMENT

This chapter examines past and possible future gaseous emissions from the transport sector and their contribution to global warming. Other environmental effects of transport are also discussed. Emissions from the transport sector in the OECD are compared with those from other final energy use sectors. Future emission levels are discussed using a scenario generated by the IEA's World Energy Outlook model (IEA, 1991c; an updated outlook is in preparation).

2.1 TRANSPORT ENERGY CONSUMPTION

Energy consumption in the transport sector was a concern of governments through most of the 1970s and 1980s. As **Figure 2.1** shows, oil use in transport has risen sharply in all three regions of the OECD. In 1990, oil represented 43% of the OECD's total primary energy supply (TPES). The transport sector accounted for 60% of final consumption of oil products, and the share has been increasing (see **Figure 2.2**). Transport energy use grew 2.1% a year in the last decade — faster than any other sector.

Road vehicles are responsible for 82% of OECD transport energy use. Air transport consumes a further 13%, with only 5% used by railways and shipping. Road transport energy use has continued to grow steadily since 1973, at an average annual rate of 2%. Only the air sector has had a higher growth rate, averaging 2.4% since 1973.

Oil provided 92% of transport energy use in 1960, and the share had risen to over 99% in 1990 despite many governments' efforts to encourage substitution of other fuels. There has been some substitution between petroleum products, slightly offsetting the increase in total energy use by the transport sector. Diesel fuel is an adequate substitute for gasoline, and diesel engines consume less energy because of their higher efficiency. Diesel accounted for about 16% of the fuel consumed in road transport in the OECD in 1973, but by 1989 the share had grown to about 26%. This rise reflects increasing road freight transport and more use of light-duty diesel vehicles in the OECD Europe and Pacific regions. The consumption of alternative fuels, such as natural gas, within the road sector continues to be extremely small.

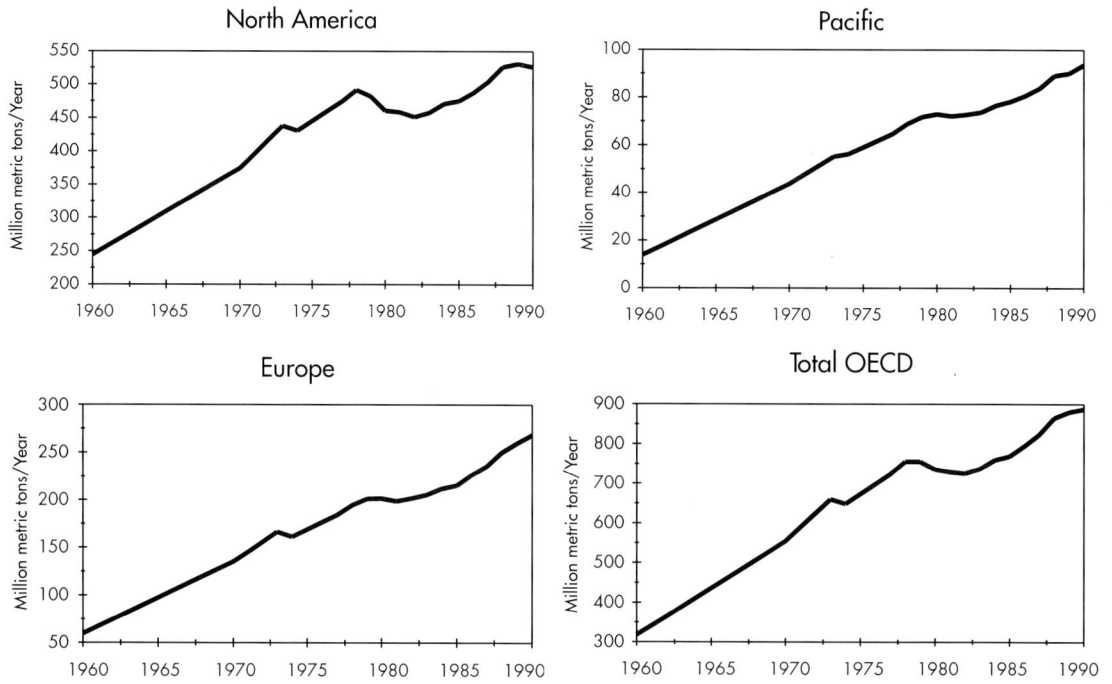

Figure 2.1
Transport Sector Oil Use in OECD Regions, 1960 to 1990
(million metric tons/year)

Source: IEA Energy Balances.

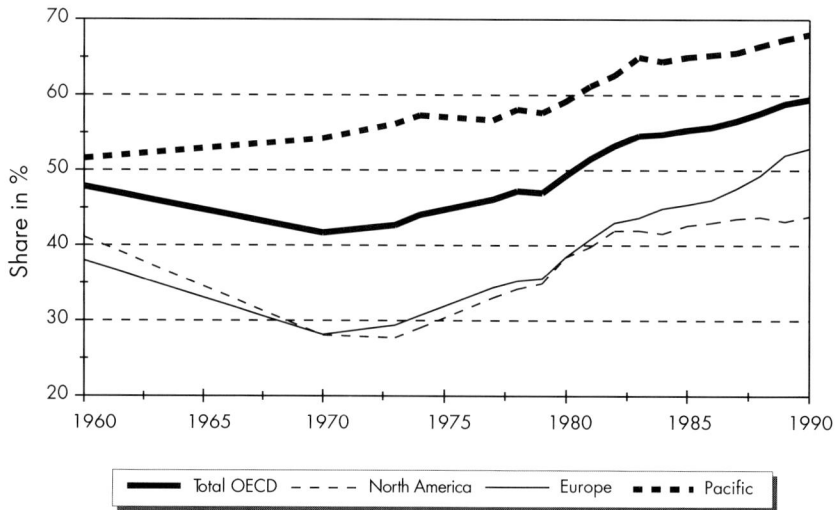

Figure 2.2
Transport Share of Total Oil Use in OECD Regions, 1960 to 1990
(%)

Source: IEA Energy Balances.

2.2   TRANSPORT AND THE ENVIRONMENT

With the increasing importance attached to quality of life and concern over rising pollution levels, transport has become a focus of environmental debate and legislation in the last two decades. Internal combustion engines used in transport generate more air pollution per unit of energy consumed than any other widely applied energy technology.

Due to its high share in total transport energy consumption, road transport is an issue of particular concern. It produces emissions of carbon monoxide, volatile organic compounds (VOCs) and nitrogen oxides (NO_x) in large volumes, having effects on local, regional and global levels. These three gases have direct consequences for health and indirect effects on atmospheric levels of ozone and methane. Ozone can form over several hours or days following emission of the gases and may affect people, animals and crops far from the pollution source. Long-range pollution of this type is common in OECD Europe. The transport sector is also responsible for a number of highly toxic air pollutants — relatively low-volume emissions that cause concern primarily because of their health effects. Most important among these are airborne lead, aldehydes and polyaromatic hydrocarbons.

Annex 1 describes health effects of transport emissions. These effects are exacerbated by the fact that high traffic density coincides with high population concentrations. While health effects are not the focus of this study, concern about local pollution is the main influence on vehicle emission standards at present. The importance of this concern in the transport market is considered in **Chapter 11**.

Transport Contribution to OECD Greenhouse Gas Emissions. The transport sector produces several greenhouse gases. They include CO_2, methane and nitrous oxide (N_2O). Of these, CO2 contributes most to the greenhouse impact of the transport sector. Some other pollutants produced by the sector — carbon monoxide, VOCs and NO_x — are "indirect" greenhouse gases, affecting the atmospheric concentrations of several "direct" greenhouse gases. The impact of both kinds of greenhouse gases can be calculated by converting to equivalent units of CO_2 emissions, using "global warming potentials" or GWPs. **Annex 1** outlines the current state of scientific understanding of these phenomena.

Table 2.1 shows the GWPs adopted for this study. They differ from those originally published by the Intergovernmental Panel on Climate Change (IPCC, 1990) and represent best estimates based on the available information. They indicate the anticipated global warming impact per unit mass emission of each gas, relative to a unit mass emission of CO_2 over the 100 years following the emission. The GWP of NO_x is particularly uncertain. Computer models of atmospheric chemistry indicate that NO_x emissions will result in increases in atmospheric ozone but decreases in methane. The overall global warming impact could be positive or negative, but it is expected to be quite small. A value of zero has therefore been adopted for calculations in this report. For carbon monoxide, methane and other hydrocarbons, the GWPs are known to be positive and the values presented here are believed to be of the correct order of magnitude. Future research, however, may result in revisions in the estimates of 20% or more.

The greenhouse gases considered in this chapter are CO_2, methane, carbon monoxide and NO_x. Emission factors for these gases across all energy end-use sectors were developed for various regions in an IEA study (IEA, 1991d). Calculations in that study are based on information on the

Table 2.1
Global Warming Potentials,
Including Indirect Effects

Gas	Estimated Lifetime, Years	Global Warming Potential Time Horizon, Years		
		20	100	500
CO_2	about 120	1	1	1
Methane	10.5	71	26	11
N_2O	132	250	270	170
Carbon monoxide	-	7	3	2
NO_x	-	0	0	0
VOC	-	31	11	6

Note: Indirect effects on methane and stratospheric ozone are excluded.

Source: Adapted from IPCC (1990, 1992).

penetration of emission abatement equipment and other technology in transport and other end-use sectors. Some of the results of the study are summarised in **Figure 2.3**, which compares approximate emissions from transport in the OECD with total OECD emissions and world-wide transport emissions.

Figure 2.3
Emissions of Greenhouse Gases in 1986: OECD Transport, OECD Total and World Transport
(million metric tons)

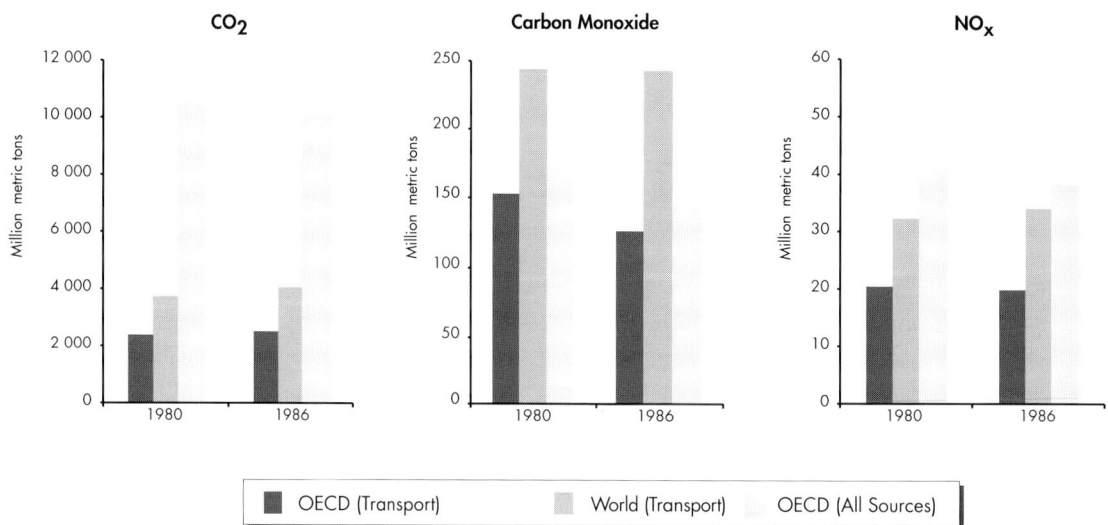

Source: IEA (1991d).

From 1973 to 1989, the increase in OECD transport-related emissions of the greenhouse gases considered in this chapter was 794 million metric tons of CO_2 equivalent. The transport contribution to the total of these emissions increased from 26% to more than 31% in the same period.

30

Carbon Dioxide. The transport sector is responsible for a rising share of CO_2 emissions from fossil fuel use in the OECD. Total annual energy-related emissions of CO_2 in the OECD rose by 7% from 1973 to 1989. At the same time CO_2 emissions from the transport sector rose 34%. In 1990 the transport share (including emissions from marine bunker fuel) was 30% (see **Figure 2.4**).

Figure 2.4
Carbon Emissions by Sector, OECD, 1990
(million metric tons)

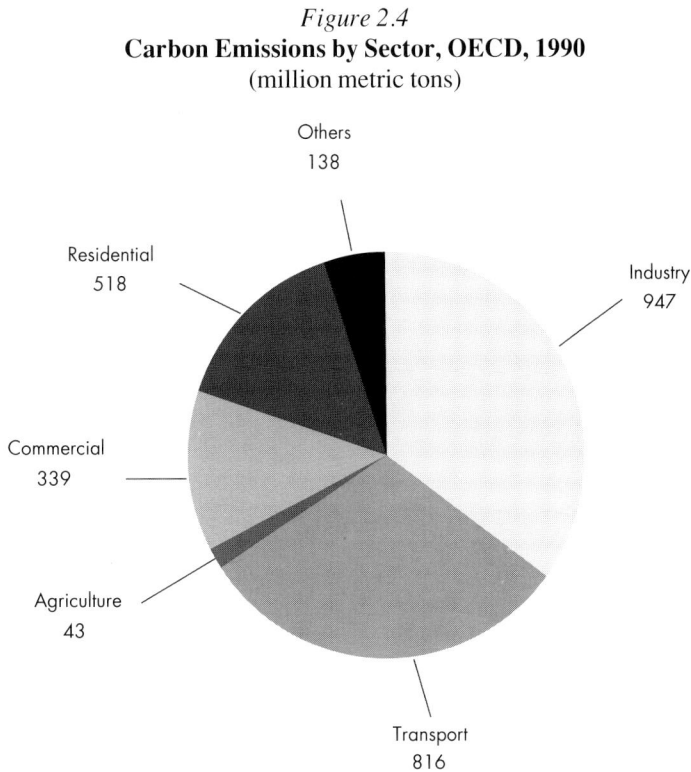

Others
138

Residential
518

Industry
947

Commercial
339

Agriculture
43

Transport
816

Source: Based on IEA Energy Balances.

Transport emissions of CO_2 can be broken down by mode: road, rail, air, inland navigation and international shipping. This breakdown is shown for 1990 in **Figure 2.5**. The data used for the graph take account of upstream emissions from oil refining, power generation and other conversion and transmission processes. Road vehicle fuel consumption accounts for roughly 75% of transport CO_2 emissions. Air transport contributes 12%, inland and marine waterborne transport 9% and rail 3%.

Carbon Monoxide. Nearly 90% of carbon monoxide emissions in OECD countries come from light-duty gasoline vehicles. With growing use of catalytic converters, these emissions have been declining.

Nitrogen Oxides. Gasoline vehicles are responsible for over 50% of the NO_x emitted from combustion of fossil fuels in the OECD. Again, emissions from cars are declining as a result of the increasing use of catalytic converters. Growth in goods vehicle traffic is expected to result in an upturn in road transport NO_x emissions in Europe early in the 21st century.

Figure 2.5
Carbon Emissions from Transport Oil Use, OECD, 1990
(million metric tons)

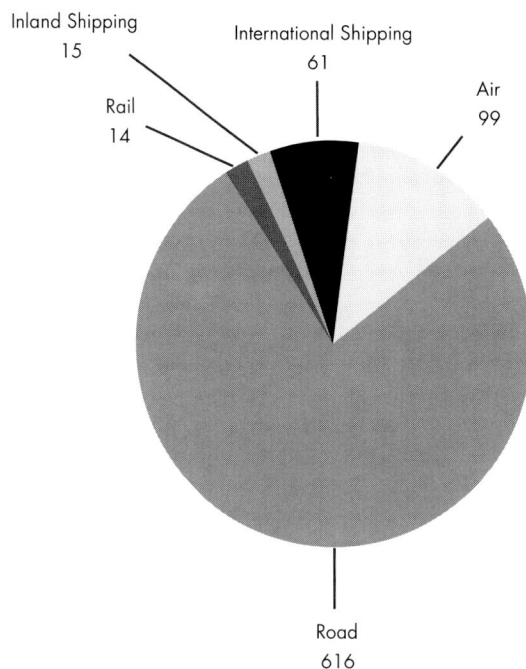

Inland Shipping
15

International Shipping
61

Air
99

Rail
14

Road
616

Source: Based on IEA Energy Balances.

2.3   TRANSPORT ENERGY OUTLOOK

Energy requirements within the OECD are portrayed up to 2005 in the IEA's World Energy Outlook (IEA, 1991c). The projections are based on the assumption that government policies affecting energy use will remain stable. In the scenario adopted, crude oil prices settle at around $21 (US 1990) per barrel in 1992. They gradually rise to about $35 early in the next decade and remain at that level thereafter. Economic activity expands at an average rate of 2.7% a year in the OECD, 3.1% in the former USSR and 4.6% in eastern Europe.

Based on these assumptions, TPES in the OECD rises 18% from 3 984 Mtoe in 1989 to 4 689 Mtoe in 2005[1]. In the Outlook, a significant proportion of the increase is supplied by an expansion in electricity generation from nuclear plants, hydropower and natural gas. Oil supply rises by 10% and supply for coal and other solid fuels by 30% (see **Figure 2.6**).

The main increase in oil demand is due to the transport sector, which more than offsets the decline in oil demand in other sectors. Transport oil demand grows at an average rate of about 2% a year

---

1.   Not including petrochemical feedstocks and non-energy uses.

Figure 2.6
Primary Energy Supply in the OECD: History and Outlook, 1975 to 2010

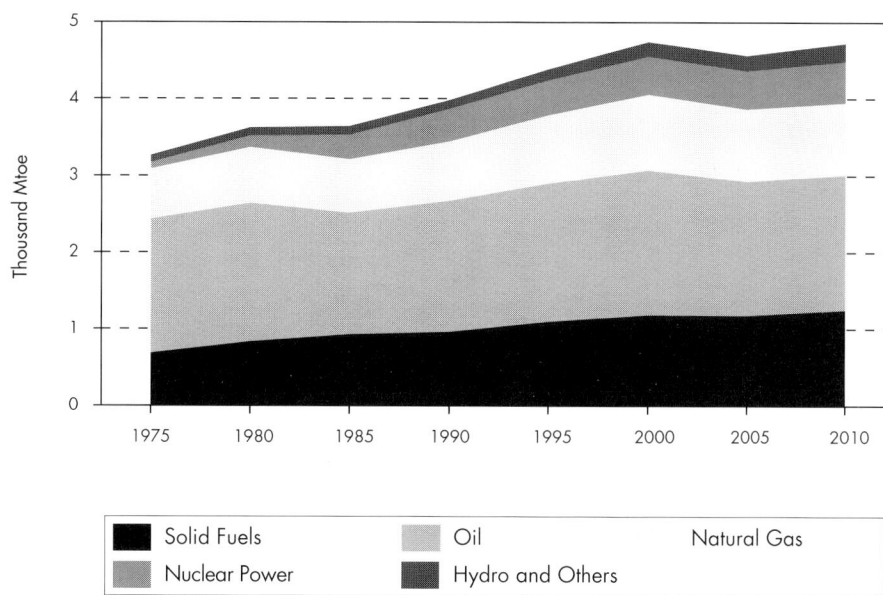

Source: IEA, National Forecasts (1991).

until 1995 and about 1.3% thereafter to 2005. The deceleration is attributable to higher oil prices in the late 1990s, a slow-down in the rate of OECD population growth and saturation in end-use demand.

All of the transport energy demand growth is in the air and road sectors. In Europe, air transport energy demand increases from around 31 Mtoe in 1990 to over 45 Mtoe in 2005. In the OECD as a whole, air transport energy demand increases from 118 to 146 Mtoe in this period. The biggest increases in road transport are for Europe and the Pacific.

Oil demand rises most in the developing countries (3.2% a year). The reasons include rapid population growth, urbanisation, greater transport needs and accelerating industrialisation. The former centrally planned economies also show strong growth in oil demand in the Outlook (2.1% a year). Growth in oil demand is slowest in OECD countries (0.8% a year). As a result, the non-OECD countries' share of world oil consumption rises from 39% in 1990 to about 52% in 2005.

Greenhouse Gas Emissions. The World Energy Outlook shows a 22% increase in emissions of CO_2 from the OECD, from 10.2 billion metric tons in 1990 to 12.5 billion in 2005. Estimates for non-OECD countries show an increase in total CO_2 emissions of 73% (8 billion metric tons). Developing countries, including China, increase their emissions by about 6 billion metric tons. Total world CO_2 emissions in 1990 amounted to about 21.5 billion metric tons, as shown in **Figure 2.7**. Emissions in the Outlook increase nearly 50% by 2005, to 32 billion metric tons.

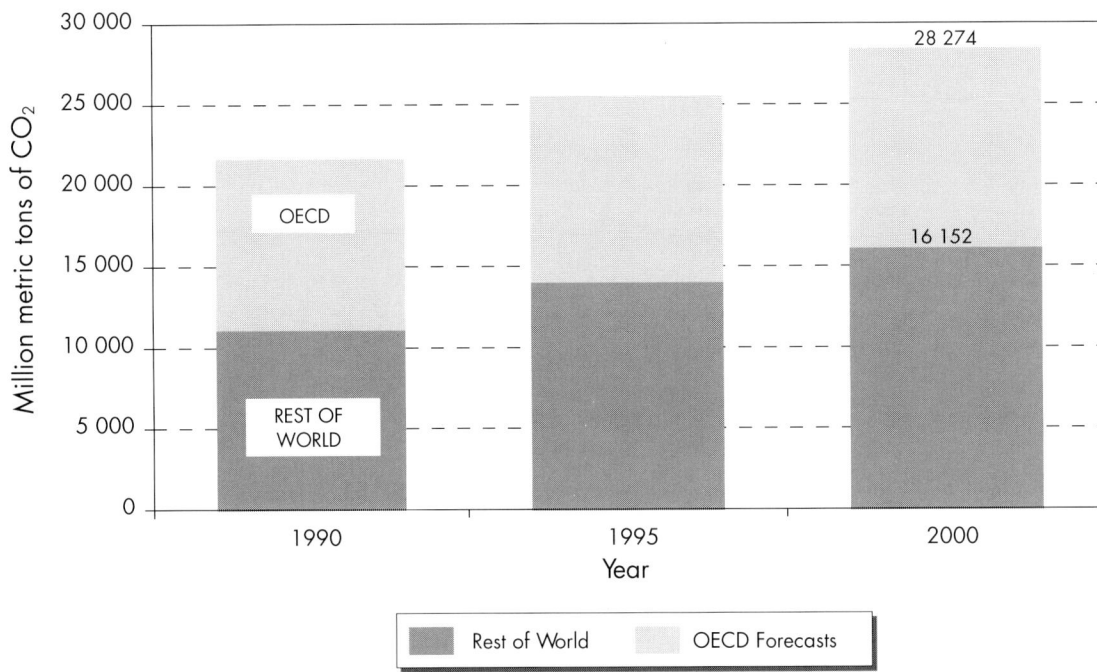

Figure 2.7
World CO$_2$ Emissions to 2000

Source: OECD emissions: IEA, National Forecasts (1991); Non-OECD emissions: IEA (1991b).

34

CHAPTER 3

GREENHOUSE GAS EMISSIONS FROM TRANSPORT: MODAL COMPARISON

This chapter compares road, rail, air and sea transport. For each mode the trends in energy use and energy intensity[1] are discussed, and estimates of greenhouse gas emissions are presented.

3.1 ROAD PASSENGER TRANSPORT

3.1.1 Cars

Car Use. Travel by car increased rapidly following World War II throughout the OECD. Passenger travel by other surface transport modes has increased more slowly or declined. As societies are increasingly shaped around car use, people travel farther to work, shop and socialise. **Figure 3.1** shows changes in passenger travel by car in six major OECD countries between 1970 and 1988. Travel grew in all six countries, with the most rapid increase in Japan, where car use more than doubled in the period. In the United States the increase was 39% (Davis and Morris, 1992).

According to travel surveys in the United Kingdom (Department of Transport, 1992) and the United States (Davis and Morris, 1992), commuter travel is growing more slowly than travel for other purposes. Cars are increasingly being used for shopping and other personal business. Traffic is growing most rapidly on highways and rural roads. In urban areas congestion has slowed the rise in traffic.

Energy Use by Cars. While cars in the United States have become more efficient in the last two decades, in Europe and Japan fuel economy has changed very little. In some countries efficiency has declined. Vehicle occupancy has declined by between 5% and 20%, depending on the country[2], resulting in higher energy use per passenger-kilometre in most countries (see **Figure 3.2**).

---

1. Energy use per passenger-kilometre for passenger modes, per metric ton-kilometre for freight modes.
2. Vehicle occupancy data are generally based on roadside observations. The quality of data is likely to vary between countries.

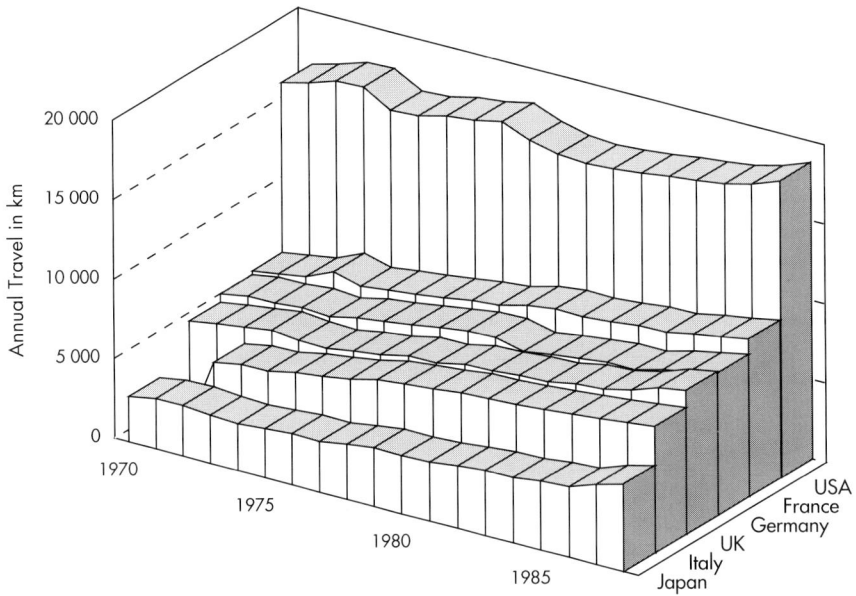

Figure 3.1
Annual Per Capita Car Travel, 1970 to 1988
(km)

Source: Schipper (1992).

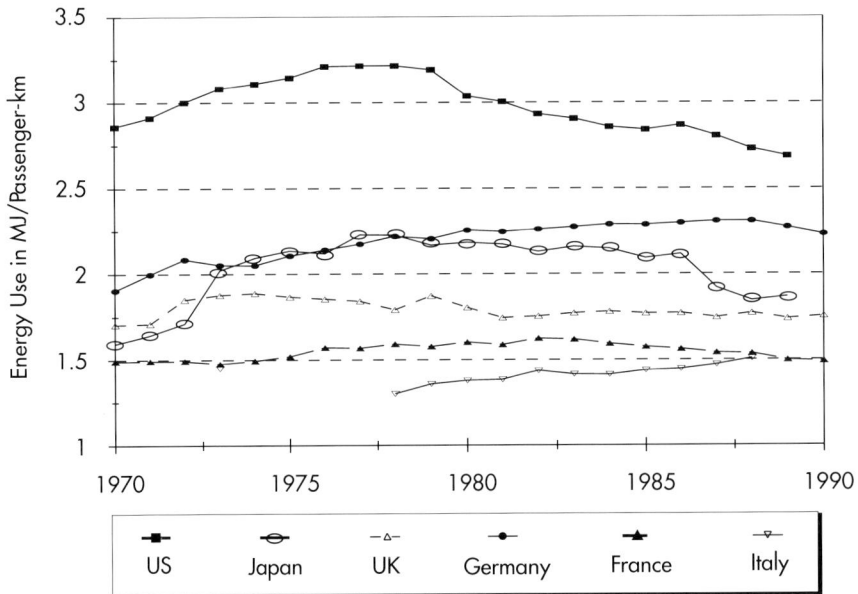

Figure 3.2
Car Energy Use per Passenger-km, 1970 to 1990
(MJ)

Source: Schipper (1992).

Gasoline fuel use for cars in the OECD is in the region of 8-12 litres/100 km or 3-4 MJ/vehicle km. With typical occupancy of 1.5-1.9 passengers per vehicle, the energy intensity of this mode is around 1.5-2.5 MJ/passenger-km.

Emissions. Manufacturing accounts for around a tenth of the life-cycle greenhouse gas emissions associated with a car with a useful life of 175 000 km. A further 15-20% is due to fuel extraction, processing and transport. Tailpipe emissions other than CO_2 contribute a further 10% to the greenhouse impact of car use. Hence tailpipe CO_2 emissions account for 60-65% of the life-cycle emissions considered here.

Local pollution due to cars continues to cause problems, especially in urban areas, despite the imposition of emission standards requiring catalytic converters to be used. Carbon monoxide and VOCs are the main pollutants from cars with catalytic converters. They are produced by cold engines before the catalysts are hot enough to function.

3.1.2 Buses

Bus Use. Travel by bus has increased slightly in the OECD as a whole in the last two decades. The most significant increase took place in the United States, where bus travel rose by 2% a year between 1970 and 1987. In western Europe the trend varies by country, with a rise over the period in France, Italy and Germany and a decline in the United Kingdom (see **Figure 3.3**).

Energy Use in Buses. Energy use per passenger-kilometre depends on the number of passengers per vehicle. The energy intensity is usually about a quarter to a third as much as that for cars. In countries for which data are available there has been little change in the energy intensity of this mode in the last two decades (Davis and Morris, 1992; Martin and Shock, 1989). Vehicle design and engineering has improved, but bus weight has increased because of a combination of factors including higher safety standards and levels of comfort.

Fuel use by a bus weighing 10-12 metric tons is in the range of 30-50 litres/100 km (12-19 MJ/vehicle-km), depending mainly on passenger loading, traffic conditions and frequency of stops. Average bus occupancy ranges from 10 to 25 passengers per vehicle, depending on the country. The energy intensity of bus use is between 0.5 and 1.2 MJ/passenger-km (see **Figure 3.4**).

Emissions. Buses are generally powered by direct injection diesel engines. Non-CO_2 greenhouse gas emissions from these engines are very low. Emissions associated with fuel production contribute about 10% to life-cycle emissions. Vehicle manufacture contributes a further 5-10% to overall emissions associated with bus use.

Local pollution can be a problem with urban buses and has led to the imposition of special emission standards in some countries. The pollutants that cause most concern are particulates and NO_x. Standards coming into force during the 1990s will reduce particulate emissions to levels that are barely measurable and will roughly halve emissions of NO_x from buses in OECD countries.

Figure 3.3
Inland Public Transport Travel per Capita, by Mode
(km)

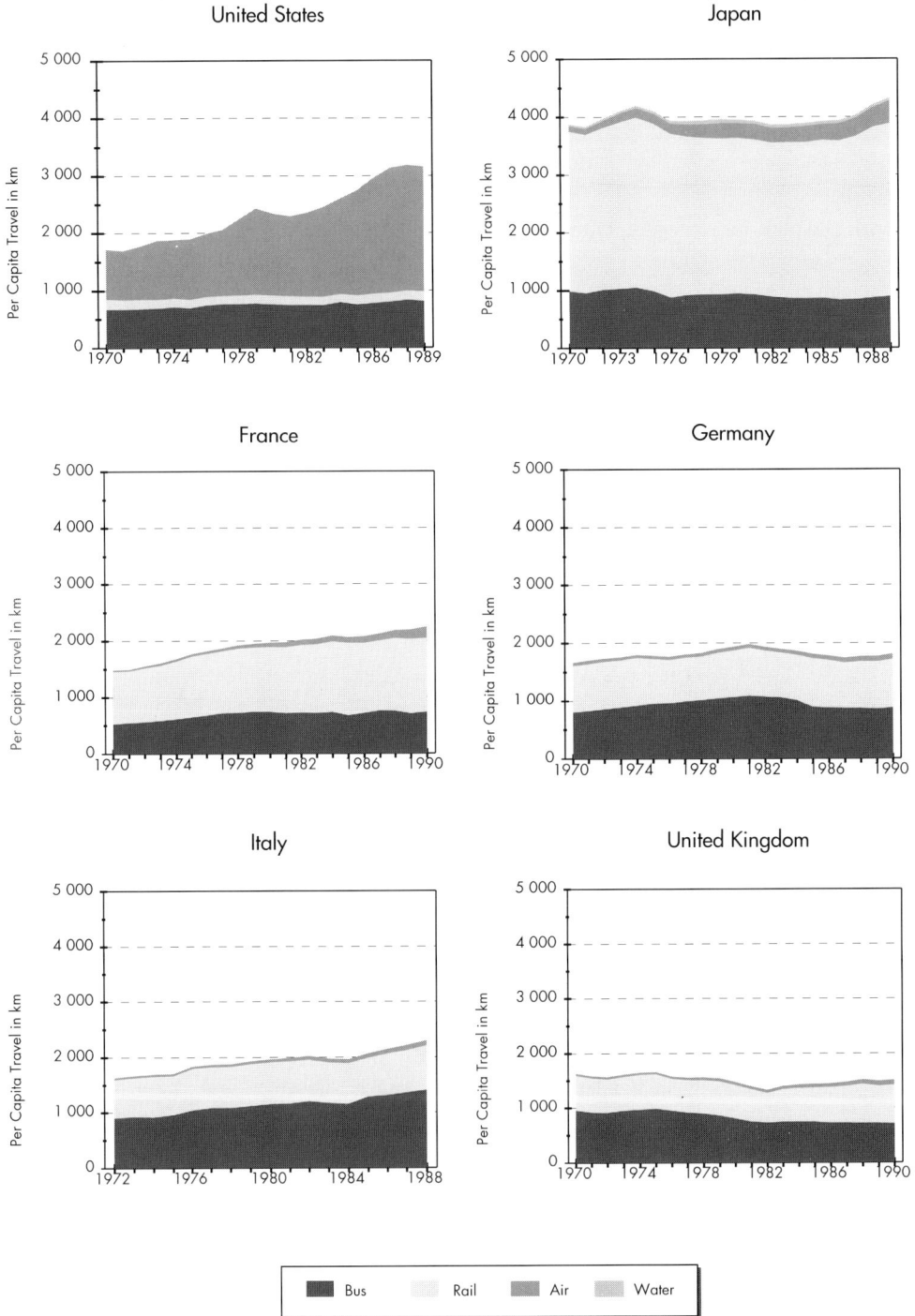

United States

Japan

France

Germany

Italy

United Kingdom

Bus   Rail   Air   Water

Source: Schipper (1992).

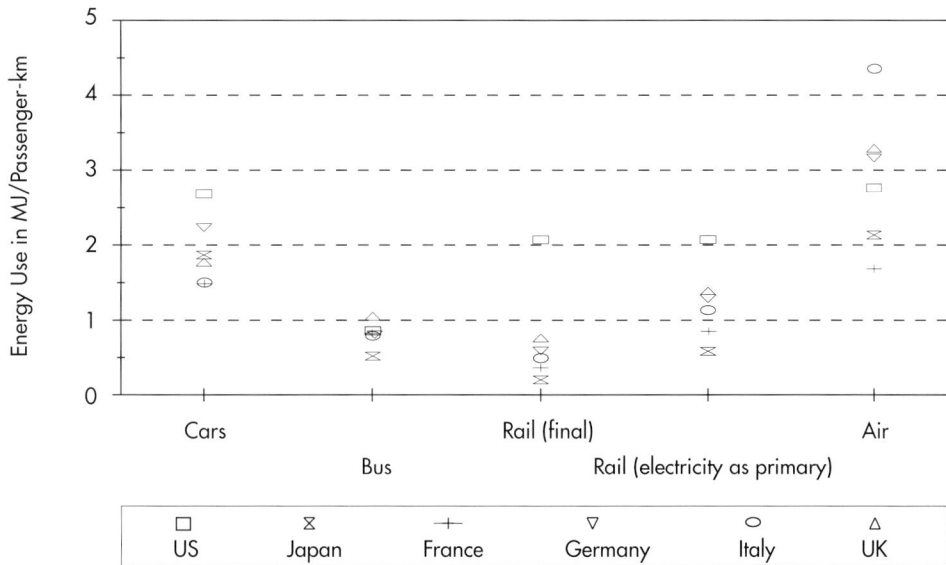

Figure 3.4
National Averages of Energy Use, by Inland Transport Mode
(MJ/passenger-km)

Source: Schipper (1992).

3.2   ROAD FREIGHT TRANSPORT

Goods Vehicle Use. From 1970 to 1989, goods vehicle traffic in the OECD is estimated to have grown from 666 billion vehicle-km to 1 588 billion vehicle-km (OECD, 1991b). This represents a mean growth rate of 4.7% a year. The largest increase has been in the use of light goods vehicles. At the same time, heavy goods vehicles have tended to increase in size and to take an increasing share of the bulk freight market. The volume and modal split of freight traffic is shown for six countries in **Figure 3.5**.

Goods Vehicle Energy Use. Several competing trends affect the energy intensity of road goods transport. The increasing number of light vehicles tends to increase energy intensity, while the increasing size of heavy vehicles results in lower energy use per ton-kilometre. Engine efficiency has been improving, with increasing use of turbocharging, intercooling and other advances. The ratio of engine power to vehicle weight is also rising, however, and this tends to result in higher energy use.

Detailed information on energy intensity is limited, although some estimates for individual countries are available. In the United Kingdom, road freight energy intensity increased from 3.4 MJ/ton-km in 1980 to 3.6 MJ/ton-km in 1986 (Martin and Shock, 1989).

Considerable potential remains for energy efficiency improvements in road freight transport. Aerodynamic styling of vehicles is beginning to improve. It can cut air resistance by a factor of

Figure 3.5
Freight Traffic and Modal Shares
(metric ton-km/capita)

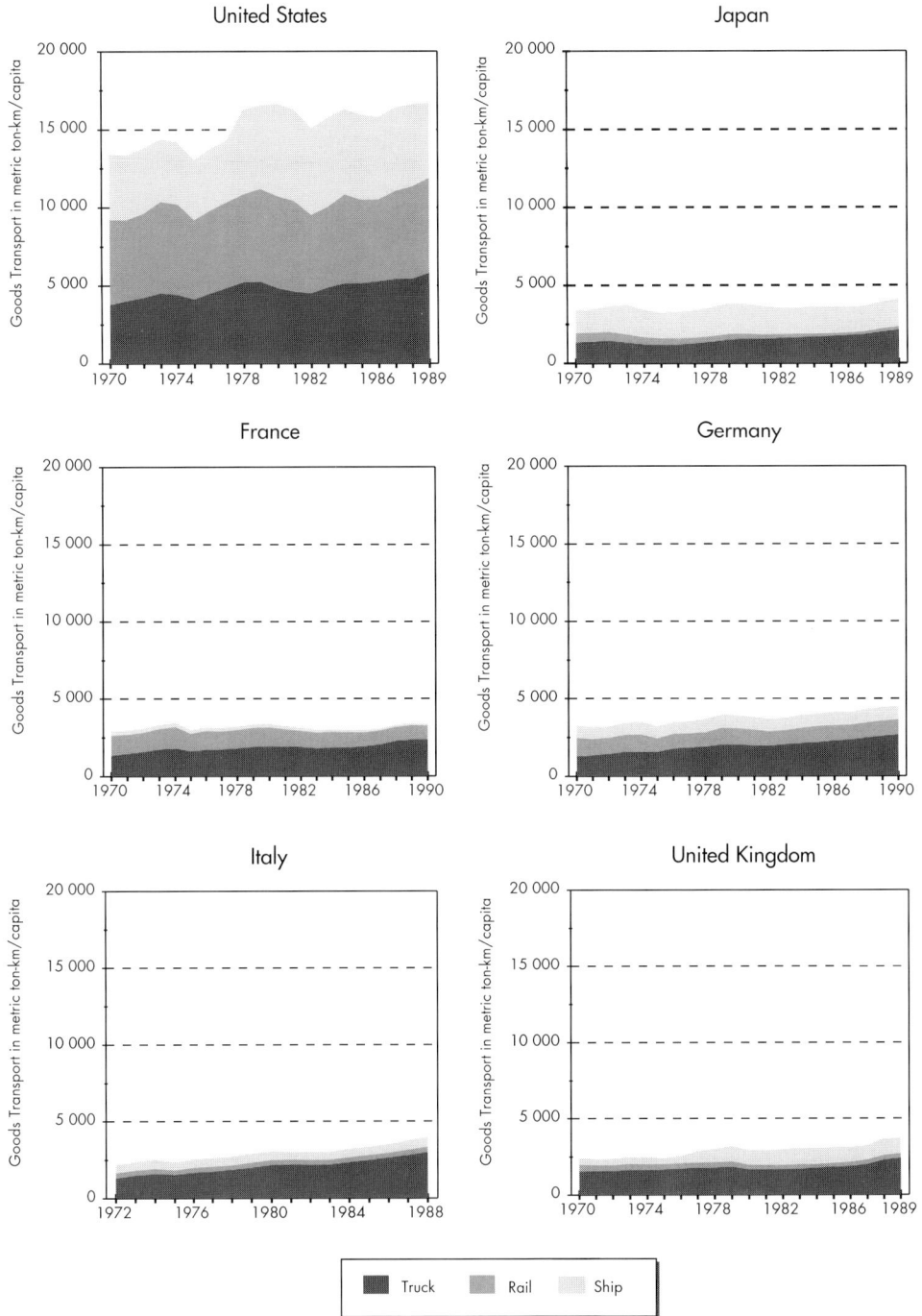

United States

Japan

France

Germany

Italy

United Kingdom

Truck   Rail   Ship

Source: Schipper (1992).

two in many cases, and reduce energy use by about 20%. Vehicle weight reduction can save energy by reducing the effort required to move the vehicle or by allowing a larger load to be carried without exceeding weight limits. Engines and transmissions continue to be improved.

Emissions. Most road freight transport is carried by heavy duty vehicles with direct injection diesel engines. Greenhouse gas emissions from these engines are dominated by CO_2. Emissions from the rest of the fuel cycle and vehicle manufacture together form 20% or less of overall emissions.

Local pollution from road freight is significant and growing rapidly, causing concern mainly in urban areas. As in the case of buses, the most important pollutants are particulates and NO_x.

3.3   RAIL PASSENGER TRANSPORT

Train Use. Rail travel has increased in most OECD countries over the last two decades. Railways have increasingly concentrated on markets where the volume of traffic justifies the high cost of constructing and maintaining rail infrastructure. The main passenger markets are intercity and commuter travel. Rail services for rural areas and small towns have declined or been withdrawn.

Energy Use in Trains. In the OECD as a whole, about 28% of final energy use by rail transport in 1990 was electricity, and most of the remainder was diesel fuel. Primary energy use does not differ significantly between electric and diesel trains. The energy intensity of rail travel has improved for some applications and worsened for others. Energy use per passenger-kilometre is affected by the interplay of several factors, including occupancy, tractive unit efficiency, aerodynamic design and train weight.

Train seat occupancy rates vary widely between types of service and between regions. Occupancy is one of the main determinants of energy use per passenger-kilometre. In general the highest occupancy is achieved by intercity services. Some services, such as the French Trains à Grande Vitesse (TGVs) achieve occupancy levels in the region of 70%. Occupancy of about 40% is more normal, resulting in primary energy intensity of 1-1.5 MJ/passenger-km.

Traction units continue to improve in efficiency. In diesel locomotives small energy savings are possible using more advanced engines and transmissions. In electric locomotives both electric motors and power electronics are becoming more efficient.

Aerodynamic styling of trains is also improving. Whereas intercity trains designed in the 1960s had drag coefficients of around 3.5 or more, the more recently developed high-speed trains have drag coefficients below 2. Aerodynamic drag is the main determinant of train energy use at constant high speed. On typical rail routes, however, the speed cannot be held constant. Mass becomes a much more significant factor when frequent acceleration and deceleration at high speed are required.

In intercity services, train weight has risen in the last two decades as comfort and safety levels have increased. In the United Kingdom, for example, railway carriages built in the 1950s weighed around 30-33 metric tons, but new carriages in 1990 weighed around 40 metric tons. This pattern has been followed in other OECD countries. In Germany, intercity rail carriages now weigh close to 50 metric tons. The weight rise is due mainly to the addition of air conditioning, automatic doors, sealed toilet systems and higher furnishing standards. The effect of increasing train weight has tended to balance the effect of reduced drag, so that energy use per train-kilometre has not varied much.

Emissions. Electric rail traction is important in Europe and Japan. North America and Australia rely mainly on diesel traction. Most countries that use electric traction also have a high level of non-fossil power generation. As a result, electric traction produces lower greenhouse gas emissions in these countries than would arise from diesel traction.

In countries with a high proportion of fossil fuel use in power generation, electric traction produces higher greenhouse emissions than diesel. This is illustrated in **Figure 3.6**, which compares the greenhouse gas emissions resulting from the provision of 1 kWh of useful work using different types of traction.

Figure 3.6
Greenhouse Gas Emissions from Tractive Effort Provision
(grams of CO_2 equivalent per kWh)

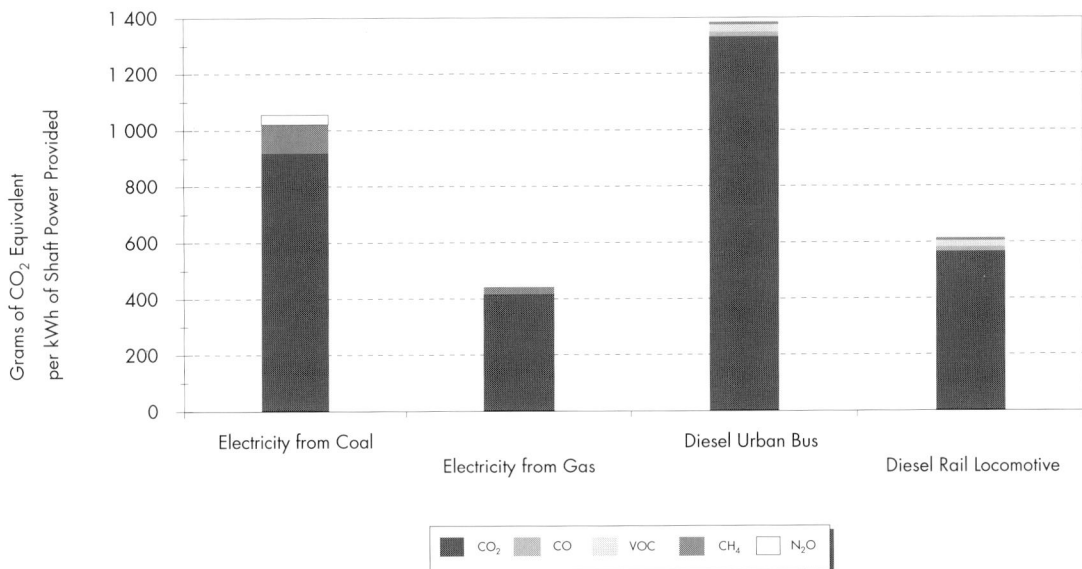

Source: CEC, (1992).

Diesel train emissions are not regulated in most countries. However, members of the International Union of Railways have agreed voluntary emission standards for railway engines (see **Table 3.1**). Diesel engines designed for optimum energy efficiency can meet these standards with little difficulty. The main local pollutants produced are NO_x and particulates (which are not covered by the standards), but as most of the pollution occurs some distance from concentrations of people, it causes less concern than road vehicle emissions.

Table 3.1
Emission Standards for Rail Locomotive Engines
Average Emissions on Selected Points of the ISO 8159 Engine Test Cycle
(g/kWh engine output)

	NO$_x$	Carbon Monoxide	VOC
Engine Entering Service:			
Until 1982	24	12	4
From 1982	20	8	2.4
From 1993	16	4	1.6

Source: British Rail (1993).

3.4   RAIL FREIGHT TRANSPORT

Rail Freight Traffic. Overall rail freight volume has declined in the past two decades, although there are significant differences between OECD countries. Road transport is more flexible and has falling relative costs, so traffic has been attracted away from the railways. Nevertheless, rail freight remains very significant in the United States and continental Europe. Rail is the preferred mode of transport for minerals, construction materials and heavy manufactured goods.

Energy Use. The energy intensity of rail freight has been reduced, partly through the narrowing of the market to trainload freight. An efficient trainload freight operation can achieve energy intensity of 0.4-0.5 MJ/ton-km (Rigaud, 1989). Container or mixed freight energy intensity is in the region of 0.9-1.0 MJ/ton-km.

The technical scope for energy savings in rail freight could be significant. Drag coefficients of badly marshalled mixed freight trains can be as high as 15, but better management and design changes could lead to reductions by a factor of three, with 20-30% energy savings (Gawthorpe, 1983). Improved bogie designs can reduce rolling resistance as well as track wear and noise, again offering potential energy savings of up to 20% (CEC, 1992).

Emissions. As in the case of passenger rail transport, greenhouse gas emissions are determined by the type of traction used. Diesel traction tends to be used more for freight than for passenger transport. Local pollutants produced in this case are comparable with those from road freight transport, although in smaller quantities because the energy consumption is lower. Locomotive standards are not likely to keep pace with the expected tightening of heavy-duty road vehicle emission standards in the next decade. Rail vehicles last about 30 years — three to four times as long as trucks. The lower turnover rate means that, even were standards to be imposed at the same time as for trucks, they would take longer to have an effect.

3.5   RAIL TRANSPORT CASE STUDY

Annex 2 presents a case study provided to the IEA by the Swiss Government, considering the potential environmental impact and benefit of a Swiss proposal to establish an integrated high-speed transalpine rail network. The project is not expected to be completed before 2010, when it is projected to reduce day-time car traffic on the Brenner highway by 8%, and freight traffic by about 70%. A considerable effect on greenhouse gas emissions is expected as the rail network will be powered by electricity produced largely from nuclear and hydroelectric power plants. Emissions due to passenger and goods traffic on the rail system are expected to be one-fifteenth of those that would have been produced if the traffic had gone by road.

3.6   PASSENGER TRANSPORT BY AIR

Air Travel. Demand for air travel has grown more rapidly than that for any other mode of transport in recent decades (see **Figure 3.7**). The growth has been driven by a combination of rising incomes and expectations, and falling real costs of air travel. Passenger aircraft on long haul flights are the cheapest mode of transport, with direct costs around a third those of surface modes. Much of the growth in the coming decades is expected to be in long haul flights. Very short haul flights are not cost-competitive with other modes, and only marginally competitive with rail on travel times from city centre to city centre. Even so, moderate growth is expected in this segment of the market as well.

Figure 3.7
World Air Travel, 1969 to 1989
(billion passenger-km)

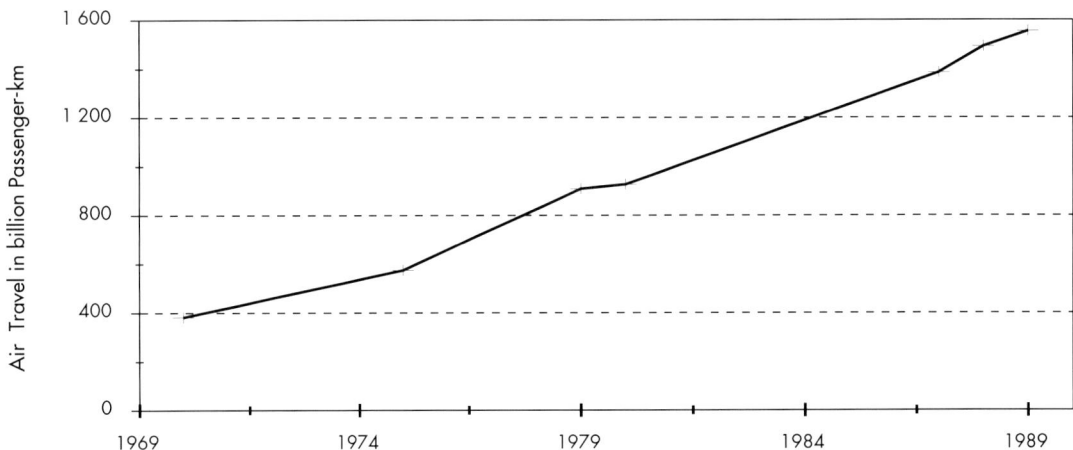

Source: *UN Statistical Yearbook* (1981 and 1988/89).

Aircraft Energy Use. Fuel accounts for 15% to 40% of airline direct operating costs, depending on the aircraft and type of service. Energy efficiency is thus one of the primary objectives in aircraft design.

Aircraft energy efficiency has improved considerably in the last three decades and this trend is continuing. Despite a 7.2% annual growth rate in world air travel during the 1980s, fuel use grew at only 3.2% a year. As a result, the share of aviation in total transport fuel demand in the OECD has changed very little, from 11.5% in 1979 to 12.9% in 1989.

World average aircraft energy consumption fell from 3.2 MJ/seat-km in 1970 to about 1.7 MJ/seat-km in 1989 (Greene, 1990). **Figure 3.8** shows how this translated into energy use per passenger-kilometre in domestic flights in the United States. Whereas occupancy in other passenger transport modes has generally declined, airlines have succeeded in increasing aircraft occupancy, using computerised booking systems and differentiated fares. The most recent generation of commercial passenger aircraft consume 1.2-1.7 MJ/seat-km and those scheduled for operation during the 1990s are expected to achieve 1.1-1.3 MJ/seat-km. Continuing efficiency improvements are being achieved through rising engine bypass ratios and pressure ratios (see **Box**). Airframes are also being improved, with increasing use of lightweight composites and improved aerodynamic design.

Box
Aircraft Engines

Most modern passenger aircraft are powered by turbofan engines. These engines have a core power unit based on a gas turbine. The turbine drives a large fan to provide most of the engine's thrust.

A turbofan engine works roughly as follows:

1. Air is drawn in at the front of the engine through a large-diameter fan.

2. Some of the air is compressed by a turbine in several stages. The rest, the "bypass" air, is ducted around the outside of the engine core.

3. Some of the hot compressed air enters the combustion chamber. Fuel is injected into the chamber, where it burns rapidly.

4. The hot combustion products are diluted by the remainder of the compressed air, reducing the temperature to 1 750°C or less.

5. The hot gases expand through a gas turbine, which provides the power for the fan and compressor turbines. There may be several successive expansion turbine stages.

6. The hot gases emerging from the turbines are mixed with the bypass air and emerge from the rear of the engine as the jet.

The efficiency of a turbofan engine is affected strongly by the bypass ratio — the proportion of air that bypasses the engine core. It is also affected by the peak pressures and temperatures achieved in the combustion chamber. The ratio between the combustion chamber pressure and that outside the engine determines the amount of work that can be obtained from the expansion turbine. The peak pressure and temperature are closely linked. They are limited mainly by the high-temperature performance of turbine blade materials.

Figure 3.8
Aircraft Energy Use per Passenger-km, US Domestic Flights, 1970 to 1989
(MJ/passenger-km)

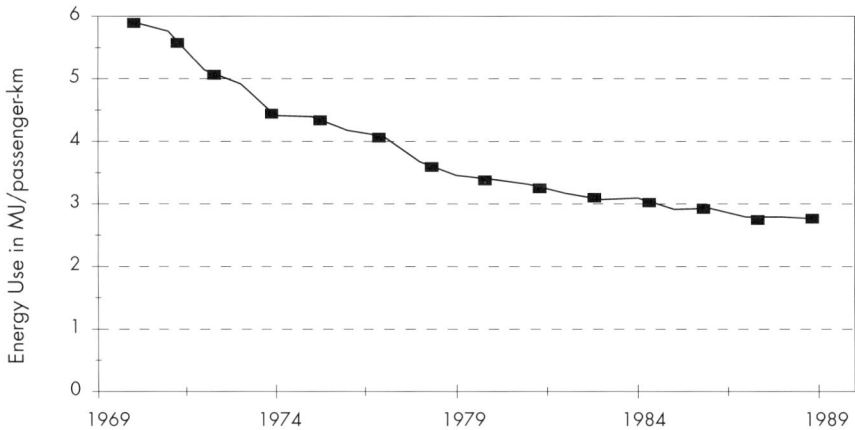

Source: Schipper (1992).

Energy Saving Potential. New aircraft technology could improve fuel economy by about 20% between 1990 and 2005 as a result of market-led technical change. Accelerated introduction of advanced technology could increase this to perhaps 30%. Other changes in air travel, including an increasing share of long-haul flights and a continued rise in aircraft occupancy, might result in a further 10% reduction in energy use per passenger-kilometre.

Air Transport Emissions. Aircraft engines have much lower emissions of carbon monoxide, VOCs, NO_x and particulates, per unit of energy consumption, than road vehicle engines. International emission standards are set for these pollutants by the International Civil Aviation Organisation based on emissions in a standard landing and take-off cycle. New engine designs have to be certified in individual countries. Certification data are available for all existing commercial aircraft engines, although the correlation with actual emissions is uncertain.

Environmental concern has focused mainly on noise and smoke produced by aircraft taking off. Early low-bypass engines were particularly noisy and dirty, causing complaints from residents near airports. New aircraft engines have far less impact, although noise remains a problem for city airports.

Recently, concerns have been raised regarding aircraft NO_x emissions. There are two possible global environmental consequences of NO_x emissions at high altitude. In the stratosphere, at altitudes of around 25 km, NO_x takes part in reactions that destroy ozone, exacerbating depletion due to CFCs. Many people in the aviation industry see supersonic stratospheric aircraft as the next major technological thrust for air transport. If such aircraft are to be acceptable, they will need engines that produce NO_x at perhaps a quarter or less of the rate of current engines[1].

---

1. NO_x and other aircraft emissions are almost always discussed in terms of grams of pollutant per kilogram of fuel. Current aircraft NO_x emissions are in the region of 10-20 g/kg. The target levels for advanced engines for supersonic stratospheric aircraft are about 3 g/kg.

46

The other source of concern about NO_x relates to its impact in the upper troposphere. This is the coldest region of the atmosphere. The low temperature affects the way emissions interact with the atmosphere and with sunlight. Models of atmospheric chemistry indicate that emissions close to the tropopause (10-12 km) are likely to result in significantly raised levels of ozone (Johnson and Henshaw, 1991). At this height, higher ozone levels are likely to have a strong greenhouse impact. The low temperature enhances the net absorption of infrared radiation. This is the height at which aircraft cruise, and therefore at which most of their emissions are produced. Estimates of the greenhouse impact of aircraft NO_x emissions indicate that it could be as great as that of their CO_2 emissions (CEC, 1992).

There is currently no empirical evidence for the effect of aircraft NO_x emissions on high-altitude ozone levels, though an investigation programme is under way in Europe[1]. Measurements of NO_x emissions from aircraft during cruise are planned, but none have yet been made. Emission levels have been estimated from computer models of the conditions in aircraft engines, and from measurements on test rigs. The range of estimates is considerable — from 10g to 18g of NO_x per kilogram of fuel burned (CEC, 1992).

Significant NO_x abatement could involve substantial changes in engine technology. There are technologies available that can probably reduce these emissions by about 50%. They require demonstration in civil aircraft but could be commercialised by 2005.

Another possible climatic impact of aircraft stems from the ice crystals in their contrails, which form in some weather conditions. Contrails may lead to the formation of thin, high-level clouds, and some models indicate that this can result in warming at the earth's surface (Schumann, 1990). This effect is even more uncertain than that of NO_x, but could also be as great as that due to aircraft CO_2 emissions.

3.7   AIR TRANSPORT CASE STUDY

Annex 3 presents an analysis of the opportunities for greenhouse gas abatement from aircraft, provided for the IEA by the Swiss Government. The allocation between countries of emissions due to aircraft is identified as a particular difficulty. Different methodologies can result in figures differing by a factor of four for Switzerland. It also finds that the expected increase in air traffic to and from Switzerland will far outweigh any progress in aircraft efficiency, so aviation fuel use will probably increase by about 50% between 1990 and 2005. The study identifies policy options to manage air transport demand and aircraft efficiency. It highlights the importance of international co-operation in any intervention in air transport.

3.8   FREIGHT TRANSPORT BY SHIPS

Use of Water Transport. International seaborne goods transport is driven by world trading patterns. Following a decline during the late 1970s and early 1980s, there was an increase in shipping energy use during the late 1980s (see **Figure 3.9**) associated with economic growth world-wide, and in Asia in particular.

---

1.   The "Aironox" programme, the main research project, examines NO_x emissions and ozone formation in aircraft wakes. It is a collaborative project involving groups in several countries, sponsored by the Commission of the European Communities (CEC).

Figure 3.9
Marine Bunker Fuel Use, OECD, 1960 to 1990
(million metric tons/year)

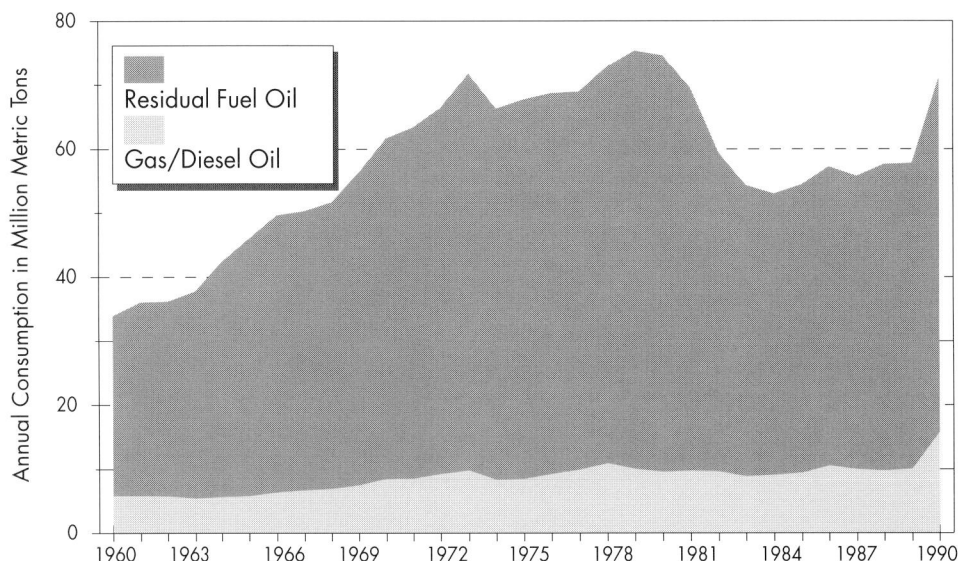

Source: IEA Energy Balances.

Coastal and inland shipping dominate freight transport in some OECD countries. Shipping is particularly important for countries with many islands or long coastlines and where transport by land is made difficult by mountainous terrain or poor infrastructure. Minerals including coal and oil are frequently moved via coastal and inland waterways. This mode of transport is cheap compared with land-based modes because of the very high capacity of individual vessels, resulting in low ratios of crew to cargo.

Energy efficiency has been a central concern in marine engine design. Motive power in most existing ships is supplied by diesel engines. In larger vessels these are generally very low-speed, two-stroke engines, with energy efficiency approaching 50%. High efficiency is obtained through:

- very high compression ratios;
- the combination of engine size and the two-stroke cycle, which minimises heat and friction losses;
- turbocharging, charge cooling and, in some cases, turbocompounding;
- constant speed operation and high load factors.

The majority of emissions arise from the main and auxiliary engines, which are either low-speed, two-stroke or medium-speed, four-stroke engines. Two-strokes are used principally on large freight vessels, four-strokes on smaller ships and ferries. Diesel fuel tends to be used in smaller, performance-oriented motors while heavy fuel oil is used in the low-speed, two-stroke motors.

48

Consumption of "international marine bunker" fuel recorded in the IEA Energy Balances represents oil delivered to seagoing ships of all flags, including warships and fishing vessels[1]. It excludes consumption by ships engaged in transport in inland and coastal waters. Inland and coastal shipping may dominate in countries with long coastlines and significant levels of marine activity (such as Norway). For the OECD on average, however, the IEA Energy Balances show that inland-water and coastal fuel consumption is roughly one quarter of marine bunker fuel consumption.

In the IEA's World Energy Outlook (IEA, 1991b), OECD consumption of marine bunker fuel continues to increase during the 1990s. However, partly because of continuing fuel economy improvements, the rise is slow, and consumption does not return to its 1973 maximum before 2005.

Marine Transport Emissions. Very little information is available on gaseous emissions from ships. In the past attention on emissions focused almost entirely on keeping smoke down to acceptable levels. Most existing emission data are either based on outdated, ill-documented analyses or have been collected from test bed engines, which are run little and in optimum conditions.

Lloyds' Register of Shipping recently undertook a major research effort to determine emission factors for different pollutants from a representative sample of the world shipping fleet under actual operating conditions (Lloyds, 1990). The emission factors calculated from this work show relatively narrow emission ranges for carbon monoxide, VOCs and CO_2, with broader ranges for NO_x (see **Table 3.2**).

Table 3.2
Emission Factors for Diesel Ship Engines in International Trade
(kg/metric ton of fuel consumed in steady state operation)

Type of Engine	NO_x	Carbon monoxide	VOC	Particulates
Medium Speed	59	8	2.7	2.5
Slow Speed	84	9	2.5	2.5

Source: Lloyds (1990).

In 1990, residual fuel oil constituted 77% of international marine bunker fuel in the OECD. This fuel often contains high levels of sulphur, nitrogen and other impurities, including toxic metals. Fossil nitrogen accounts typically for 20-30% of NO_x emissions in engines burning residual fuel oil (Lloyds, 1990). Combustion conditions and fuel quality have an impact on NO_x emissions. In general, ship engines have low emissions of carbon monoxide and VOCs but high emissions of NO_x.

---

1. The IEA Energy Balances do not include international marine bunkers within TPES but record them as a separate entry.

3.9   MARINE TRANSPORT CASE STUDY

Annex 4 presents a study commissioned for the IEA by the Norwegian Government on emissions from ships in Norwegian coastal waters. As in the case of air transport, the allocation of fuel use and emissions between countries is found to be a particular difficulty. There are also differences between the methodologies used by organisations, even within Norway, for reporting energy use by ships. This results in considerable differences in estimates of the emissions from inland shipping. The study presents a scenario for future emissions from Norwegian coastal shipping. Substantial reductions in emissions of sulphur dioxide and NO_x are expected, along with a small reduction in CO_2 emissions.

PART II

PROSPECTS FOR GREENHOUSE GAS EMISSION ABATEMENT THROUGH TECHNICAL CHANGE IN ROAD VEHICLES

CHAPTER 4

EMISSIONS FROM LIGHT-DUTY VEHICLES

Light-duty vehicles — cars, taxis, vans and light trucks — form the largest group of road vehicles in the OECD. They are responsible for most of the energy use and pollution by transport. This chapter briefly describes the technologies commonly used in gasoline and diesel engines in such vehicles, and the effects of engine parameters on emissions and efficiency. It concludes with a brief discussion of the technological potential for emission abatement in gasoline and diesel vehicles.

4.1 GASOLINE ENGINES

The majority of light-duty vehicles have gasoline engines operating on the "Otto" cycle, which is described in **Box 1**. The carburetted, naturally aspirated gasoline engine has been very successful, largely because of its low cost and high power-to-weight ratio. No alternative technology yet developed can compete on these grounds. As consumer requirements change and environmental standards tighten, it has so far proved more effective to modify the gasoline engine than to switch to other power sources.

4.1.1 Emissions

Gasoline engines have been modified to improve performance and reduce some of their negative effects, such as noise, production of smoke and emissions of gaseous pollutants. Noise and smoke have been controlled for several decades in most OECD countries. As a result, gasoline engines produce very little smoke when they are functioning normally. They remain the main source of carbon monoxide from human activities in the OECD, despite the introduction in the early 1970s of carbon monoxide emission standards. They also produce high emissions of hydrocarbons and NO_x, even though controls on hydrocarbons began to be introduced in the early 1970s and on NO_x in the 1980s. **Table 4.1** compares emissions from gasoline engines with those from other oil-burning technologies.

Table 4.1
Typical Emissions from Oil-Burning Technologies
(g/kg of fuel)

Technology	Carbon monoxide	VOC	NO$_x$	Particulates
Distillate-Fired Gas Turbine (Electric Utility)	2	0.8	9	0.7
Distillate-Fired Industrial Boiler	0.7	0.03	3	0.3
Distillate-Fired Domestic Boiler	0.7	0.4	2.5	0.4
Light-Duty Gasoline Engine (No Exhaust Control)	200-300	20-30	20-40	<0.3
Light-Duty Gasoline Engine (With Catalytic Converter)	20-60	2-6	2-10	<0.3
Light-Duty Diesel Engine (No Exhaust Control)	15-50	2-5	10-20	3

Note: Emission factors for stationary sources are based on EPA (1986).

To comply with controls on emissions of hydrocarbons, smoke and carbon monoxide, car manufacturers have had to modify engine technology. These modifications at the same time improved efficiency. However, they also tended to increase NO_x emissions. NO_x emission standards have required the use of three-way catalytic converters.

4.1.2   Influences on Emissions and Efficiency

The efficiency of gasoline engines and the emissions from them depend on a number of technical factors. These include the compression ratio, the air/fuel ratio, the air and fuel preparation, ignition and valve control, and exhaust treatment.

Compression Ratio. The efficiency of the thermodynamic cycle in a gasoline engine can be increased if its compression ratio — the extent to which the fuel-air mixture is compressed before combustion — can be increased. The compression ratio affects the expansion ratio, the extent to which the combustion products can be expanded before the exhaust valve opens. Existing gasoline engines have compression ratios of about 8-10:1. Optimum efficiency would be obtained at around 15:1. Above this level, engine friction increases and offsets the thermodynamic efficiency gain.

Higher compression ratios lead to higher temperatures in the engine and a greater tendency for it to "knock" — that is, for the fuel-air mixture to detonate spontaneously ahead of the flame front. Engine knock results in sharp pressure increases and can damage the engine. The octane number of a fuel is determined on the basis of its tendency to cause the engine to knock. Higher-octane fuels are less likely to produce knock and can be used in engines with higher compression ratios.

Air/Fuel Ratio. In Otto-cycle engines, the air/fuel ratio affects the combustion efficiency, power output and emissions. Normally, Otto-cycle engines operate with the air/fuel ratio close to "stoichiometric" — i.e. there is just the right amount of air to allow all the fuel to burn. The "equivalence ratio" (λ) is the actual air/fuel ratio over the stoichiometric ratio. For a stoichiometric mixture, $\lambda = 1$. For a lean mixture, $\lambda > 1$.

Lean mixtures allow more efficient operation, partly because the engines are able to use higher compression ratios (they are less prone to knock) and partly because the thermodynamic cycle is more efficient if a greater mass of gas is used. Engine power, on the other hand, is higher with slightly rich mixtures ($\lambda < 1$), as these maximise both energy release per unit volume in the engine and the rate of combustion, which determines the maximum possible engine speed.

Figure 4.1 shows how engine emissions and energy efficiency vary with air/fuel ratio. NO_x emissions are strongly correlated with flame temperature, which is highest for a stoichiometric mixture. As the mixture becomes leaner, emissions of carbon monoxide decrease, its conversion to CO_2 being affected by the amount of excess oxygen. Hydrocarbon emissions are high from engines using rich mixtures, where there is insufficient air to burn all the fuel. In very lean mixtures, the low flame temperature and speed result in incomplete combustion and, again, high hydrocarbon emissions.

Figure 4.1
Effect of Air/Fuel Ration on Emissions

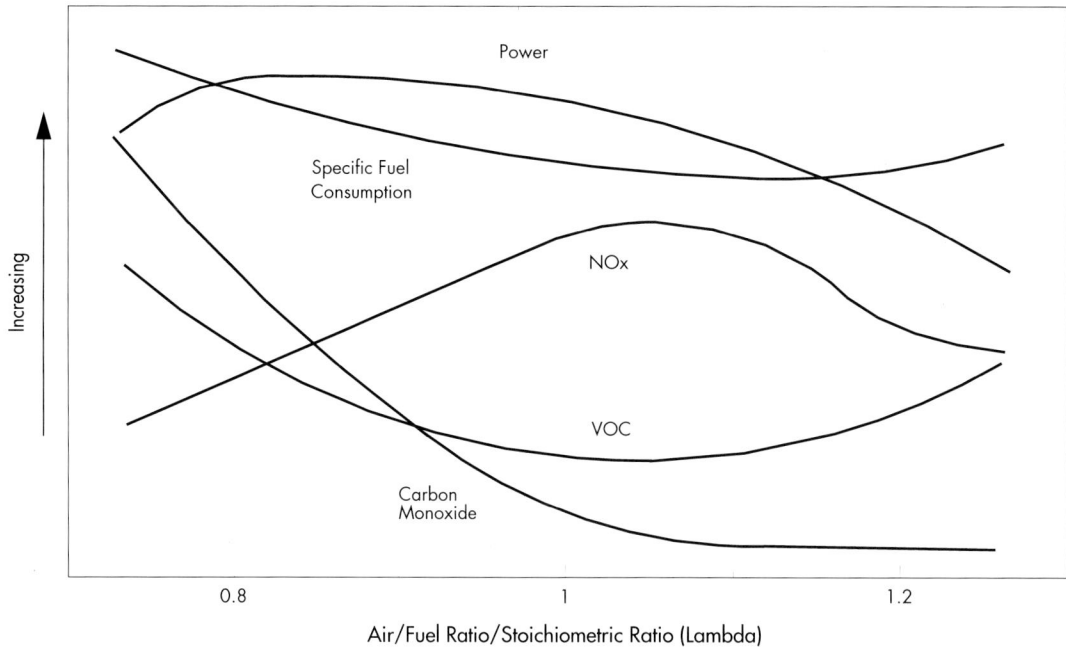

Air/Fuel Ratio/Stoichiometric Ratio (Lambda)

Air and Fuel Preparation. In a carburetted, naturally aspirated engine, the fuel/air mixture is forced into the cylinder by atmospheric pressure as the piston descends in the intake stroke. High-performance engines may be supercharged, using a turbine driven by the engine to force additional mixture into the cylinder. In a turbocharger, an additional turbine driven by the exhaust is used to power the intake turbine.

Some vehicles are available with "intercooling": intake air is first compressed, then cooled, then further compressed. This raises the density of the air and allows more to enter the cylinder. Alternatively, in "aftercooled" engines the air from the turbocharger is cooled and enters the cylinder without further compression.

In carburetted engines, the flow rate of intake air directly determines the amount of fuel added. Carburettors are carefully designed to keep the air/fuel ratio as close as possible to stoichiometric most of the time although the mixture is enriched during acceleration and cold running. A layer of fuel tends to be deposited on the intake manifold of the engine. Fuel is constantly condensing into this layer and evaporating from it, making it harder to control the air/fuel mixture entering the engine. Engine developers have done a considerable amount of work to deal with this "wetting" effect.

In fuel injection engines, intake manifold wetting is less severe, especially in multipoint injection systems where fuel can be injected directly into the open cylinder intake ports. With electronically controlled fuel injection, the air/fuel ratio can be instantaneously controlled, resulting in benefits for efficiency and emissions.

Charge Dilution Techniques. Dilution of the air/fuel mixture can reduce NO_x emissions by reducing the peak temperatures in the combustion chamber. European and Japanese manufacturers have been working on lean-burn engines, and many production models in the late 1980s used lean-burn to some extent. These engines operate slightly lean except when high power is required.

Three-way catalytic converters are effective only when the air/fuel ratio is very close to stoichiometric, and so cannot work with lean-burn engines. If the concentration of oxygen in the exhaust is significant, NO_x cannot be reduced. Emission standards in most OECD countries preclude the use of most existing lean-burn engines because of their levels of NO_x and hydrocarbon emissions. Although the hydrocarbons can be controlled with an oxidation catalyst, the NO_x emissions remain too high.

An alternative approach with many of the advantages of lean-burn is exhaust gas recirculation. This technique involves diluting the incoming air/fuel charge with exhaust gases, reducing the flame temperature and flame speed. It can be used with three-way catalytic converters as it does not raise the oxygen level in the exhaust.

Ignition. Ignition timing strongly affects engine efficiency and emissions. Ideally the air/fuel mixture should burn completely around or slightly after "top dead centre" — the point in the engine cycle when the gases are fully compressed. For this to happen, the spark-plug must fire slightly before top dead centre is reached. If ignition is too late, combustion continues late in the expansion stroke, resulting in wasted energy and incomplete combustion. NO_x emissions are lower if ignition is delayed. If ignition is too early, the piston has to compress the burning fuel/air mixture, resulting in a loss of efficiency. NO_x emissions are raised in this case because of the very high peak temperatures that result.

Modern engines increasingly use electronically timed ignition in place of mechanical distributors. This permits ignition timing to be optimised for engine conditions. At higher engine speeds the spark timing is advanced, and at low engine speeds ignition can occur very close to top dead centre. Spark timing can be optimised for fuel economy, or to minimise emissions.

Restricting combustion to a narrower period around top dead centre improves efficiency and reduces emissions. This is particularly important for engines using lean-burn or exhaust gas recirculation, where combustion is generally slower. "Fast-burn" techniques include increasing the flame speed (by increasing the level of turbulence) and reducing the distance the flame travels. Both involve changes to engine or cylinder design. Fast-burn techniques also reduce knock, allowing the use of a higher compression ratio.

The minimum spark energy for reliable ignition varies according to the air/fuel mixture and combustion conditions. It is lowest for stoichiometric mixtures, higher for lean mixtures, mixtures diluted with exhaust gases, and greater turbulence levels. Ignition delay can be reduced and variability minimised by providing a spark above the minimum energy.

Variable Valve Timing. It is possible to modify the engine compression and expansion ratios by modifying the timing of the exhaust and intake valves' opening and closing. The valves are normally operated by a rotating camshaft. One approach to variable timing is the use of a sliding camshaft with specially contoured cams. Another approach is the use of electric servos to operate the valves, allowing full electronic control. Electronic valve timing can eliminate the need for a throttle, as the air intake can be controlled by the time for which the valves are open. This gives an energy efficiency gain of about 10%. By optimising the valve timings for engine conditions, a greater energy saving can be obtained, although it is offset somewhat by the energy needed to operate the electric servos.

Catalytic Exhaust Treatment. Catalytic converters were developed to deal with the high level of unburned hydrocarbons and carbon monoxide in gasoline engine exhaust. These "two-way" or "oxidation" catalytic converters have been fitted to cars in the United States since 1975. "Three-way" or "oxidation/reduction" catalytic converters reduce emissions of hydrocarbons, carbon monoxide and NO_x, and were developed in the late 1970s. They have been required for gasoline cars in the United States since 1981. Three-way catalytic converters are now needed for gasoline cars to meet emission standards in almost all OECD countries.

The catalyst in a catalytic converter normally consists of platinum and/or palladium and rhodium. A few grams of the catalyst are distributed through a "ceramic monolith". This is an extruded ceramic block forming many fine, parallel channels through which the exhaust gases pass in close contact with the catalysts.

Catalytic converters can be "poisoned" by lead in fuel (hence the need for unleaded gasoline) as well as sulphur and phosphorous. They can also be damaged by excessive temperatures, which can arise from too much oxygen and unburned fuel in the exhaust.

Catalytic converters are designed to tolerate swings in the air/fuel ratio, provided it is stoichiometric on average. Stoichiometry is maintained most easily with electronically controlled fuel injection, using an oxygen sensor in the exhaust stream to provide feedback. This type of system, known as a closed loop catalytic converter, has become standard in OECD countries.

Evaporative Emissions. Fuel evaporation from vehicles and during refuelling and delivery can constitute a major proportion of the VOCs emitted as a result of vehicle use. Evaporative emissions from the vehicle can be reduced using canisters containing activated carbon, which absorb VOCs from the fuel tank when the vehicle is not running. When the engine is running, intake air is drawn through the canister, purging it of hydrocarbons, which then form part of the fuel in the mixture fed to the engine. In the United States and Japan, carbon canisters have been compulsory on all cars since the 1970s. They have since become compulsory in Australia, Scandinavian countries, Switzerland, Austria and the European Community (EC). VOCs may also be controlled at other stages of the gasoline fuel cycle by improving storage tank sealing and using vapour recovery units on fuel pumps.

4.2   DIESEL ENGINES

Diesel engines are increasingly being used for light-duty vehicles. They are used for a high proportion of vans and light trucks in Europe, and for cars. The majority of diesel cars have indirect fuel injection, while many commercial vehicles have direct fuel injection engines.

The diesel cycle is outlined in **Box 2.** The diesel engine, patented in 1892, has several advantages:

- It has variable stoichiometry and the air intake does not have to be throttled to limit it in proportion to the amount of fuel being used; this contributes considerably to the higher efficiency of diesel engines.

- It uses high temperature and pressure to ignite the fuel/air mixture rather than a spark, eliminating the need for some of the less reliable components of gasoline engines, such as spark-plugs.

- The fuel is injected into compressed air towards the end of the compression stroke rather than having to be drawn into the engine as a vapour. This injection approach avoids most of the cold start problems of gasoline engines.

- Compression ratios are much higher than in gasoline engines, resulting in higher efficiency.

Box 2
Diesel Compression-Ignition Engines

The four phases of diesel engine operation are as follows (as in the Otto-cycle engine, they can be carried out in two strokes instead of four):

Intake: The intake valve opens as the piston descends from the exhaust stroke, drawing in the air (more air can be added by a turbocharger) and generating turbulence to improve fuel/air mixing during combustion.

Compression: The intake valve closes as the piston rises to compress the air, increasing its temperature to between 700°C and 900°C.

Ignition: Just before the piston reaches top dead centre, fuel injectors spray diesel oil at high pressure into the cylinder; fuel injection volume is determined by power requirements. A delay occurs while some fuel evaporates and an air/fuel mixture forms and then ignites.

Power Stroke: The mixture continues to burn, maintaining the pressure in the cylinder as the piston descends.

Exhaust: The piston rises, pushing the burned gases out of the opened exhaust valve into the exhaust manifold.

Diesel engines also have a number of disadvantages. They are more expensive than gasoline engines on an equivalent power basis, and they have lower power/weight ratios. Because of the constant high compression ratio and high air intake, they are noisier than gasoline engines at low power. In cold weather heavy hydrocarbons in the diesel fuel can crystallise, causing cold starting problems.

4.2.1 Emissions

The high air/fuel ratio in diesel engines results in very low emissions of carbon monoxide and VOCs. However, because the fuel burns with a very hot flame in a stoichiometric zone as it mixes with air, it produces moderate amounts of NO_x. Emissions of NO_x from light-duty indirect injection (IDI) diesel engines are higher than those from catalyst-controlled gasoline vehicles, but lower than those from uncontrolled gasoline vehicles. Particulate emissions are much higher than those from gasoline engines, especially if the fuel is not adequately atomised and mixed with air and consequently does not burn completely.

4.2.2 Influences on Emissions and Efficiency

As with gasoline engines, diesel engines' energy efficiency and emissions depend on a number of technical factors. These include air preparation, fuel injection and catalytic exhaust treatment.

Air Preparation. To increase the power-to-weight ratio of diesel engines, many have turbocharging as a standard feature. Forcing more air to the engine cylinders allows more fuel to be burned and a smaller engine to be used for a given power level, at higher efficiency. Turbocharging results in a leaner air/fuel mixture, which results in lower particulate emissions. At high load, however, it raises the peak temperatures in the engine, yielding higher NO_x emissions.

Charge cooling (cooling of the hot compressed air between the turbocharger and the engine intake) is generally used only on heavy-duty diesel vehicles. This technique makes the air denser, so more of it can be introduced to the cylinder. The efficiency and power/weight ratio are raised further, and the combination of cooling and the leaner mixture results in lower emissions of NO_x.

Exhaust gas recirculation can reduce NO_x by a factor of two or more at moderate loads, but tends to increase particulate formation. Since the recirculation of particulates into the engine results in excessive wear, this is not likely to be an acceptable technique for diesel engines without further development.

Fuel Injection. The timing of fuel injection has an effect on engine efficiency and on particulate and NO_x emissions. Fuel is injected just before the piston reaches top dead centre. Earlier fuel injection results in higher emissions of NO_x, but lower particulates and better energy efficiency. Where NO_x is an overriding concern, it is possible to retard injection and filter out particulates from the exhaust. However, this means sacrificing much of the diesel engine's energy efficiency advantage, and it adds to the cost.

The main challenge that has been addressed by manufacturers is to achieve late fuel injection, and so low NO_x, while maintaining low particulate emissions and fuel consumption. This can be done by modifying the fuel/air mixing so that combustion occurs more rapidly and is more complete. Much of the research effort on diesel engines has focused on fuel injection systems and on combustion chamber design. Higher injection pump pressures and better injection nozzle design result in better fuel atomisation. Combustion chamber design is chosen to improve turbulence and swirl, and increase the mixing rate.

Catalytic Converters for Diesel Engines. Several manufacturers now offer diesel cars with oxidation catalytic converters. They achieve some reduction in hydrocarbon, particulate and carbon monoxide emissions. This may allow a little more freedom in fuel injection timing to reduce NO_x emissions.

A variety of options for reducing diesel NO_x emissions catalytically have been examined. However, the high oxygen level in the exhaust renders existing three-way catalytic converters ineffective for NO_x abatement. Catalytic reduction of NO_x is possible in stationary sources, using techniques involving the injection of ammonia to the exhaust stream. This requires very precise metering of ammonia to match the NO_x in the exhaust, and is not feasible in cars.

Other catalysts have been identified that will selectively reduce NO_x in an oxygen-rich exhaust stream, but they are poorly developed and have been tested only under very artificial laboratory conditions.

CHAPTER 5

THE POTENTIAL FOR IMPROVED FUEL ECONOMY

This chapter explores the technical potential for reducing fuel use by light-duty vehicles per unit of distance travelled. The first section discusses technology that can improve fuel economy for gasoline and diesel vehicles. There follows a discussion on technical change in the automotive industry and finally a section looking at the market potential for new technology in a "business-as-usual" scenario. The conclusions of this chapter provide the basis for the alternative fuels appraisal in **Chapter 6**.

Average car fuel consumption in most OECD countries is in the range 8-12 litres/100 km. Cars with fuel consumption in the range of 2.5-3 litres/100 km have been demonstrated by manufacturers (OECD, 1991b). These fuel economy levels have been achieved by improvements in aerodynamic styling, the use of lighter materials, the use of higher tyre pressures in narrower tyres and by reductions in power-to-weight ratios. The cars do not have the performance currently expected by consumers and are not being marketed.

Improved technology can be used either to increase vehicle fuel economy or to boost performance. In practice fuel economy improvements often fail to be realised as the trend is to build larger and/or more powerful cars. For example, the use of the turbocharger allows a reduction of engine size with no loss of power, resulting in a fuel economy gain. However, the turbocharger has tended to be marketed primarily as a means of obtaining more power from the same size engine, resulting in a fuel economy penalty.

Car fuel economy throughout the OECD improved between 1975 and 1991 as a result of weight reduction, improved aerodynamic design, the introduction of radial-ply tyres, the increased use of front-wheel drive, and engine efficiency improvements.

Most technical improvements in vehicle fuel economy and performance in the next decade will come from the diffusion of technology already existing in prototypes and some new cars. **Figure 5.1** outlines opportunities for fuel economy improvements in light-duty vehicles. There are two important areas where fuel economy improvements can be made: increasing the efficiency of the drivetrain (engine, transmission and axles) and reducing the work needed to move the vehicle (tractive effort).

Figure 5.1
Technical Approaches to Reducing Car Energy Use and Greenhouse Gas Emissions

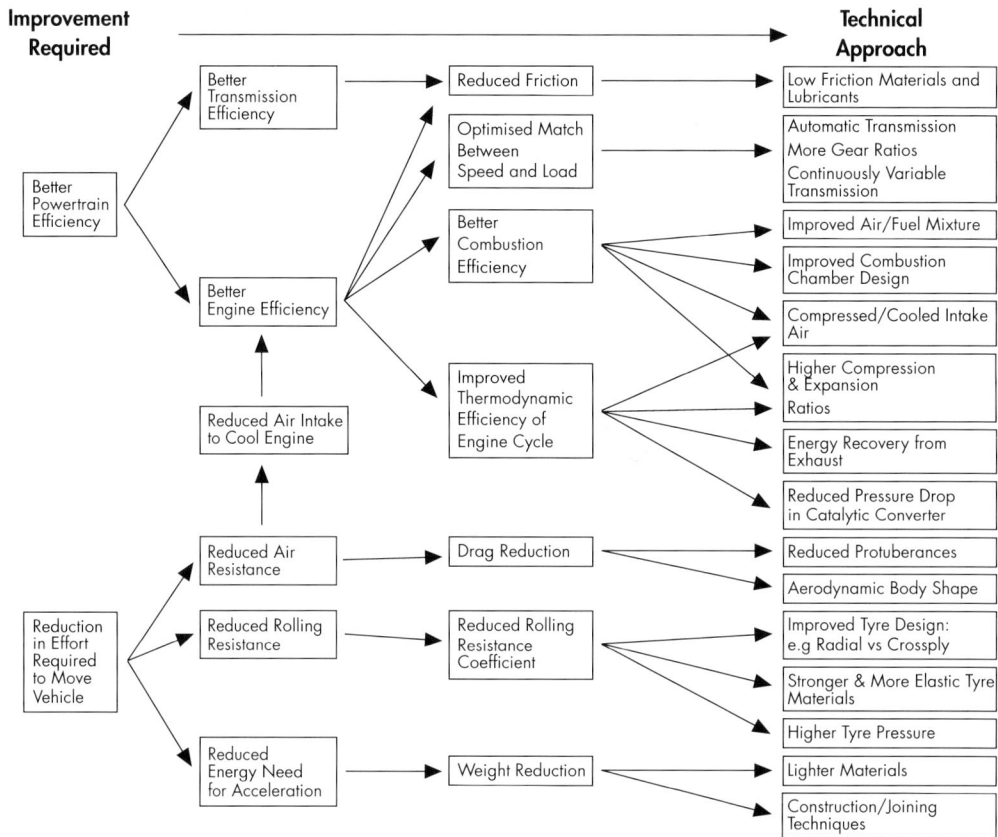

5.1   IMPROVING DRIVETRAIN EFFICIENCY

Thermal Losses to Coolant. Heat losses through cylinder walls account for about a third of the energy released in fuel combustion. Losses can be reduced by introducing air gaps or low thermal conductivity cylinder liners. Considerable effort has been devoted to developing ceramic cylinder liners, although many problems remain to be solved. Lower-conductivity cylinder walls tend to develop hot spots which cause premature ignition of the air/fuel mixture, and more heat is lost in the exhaust.

Heat losses to coolant also affect fuel economy indirectly. An increased cooling air requirement for the engine results in increased aerodynamic drag for the car. The cooling system also consumes engine power driving fans or pumps.

Energy Losses to Exhaust. In naturally aspirated engines around a third of the fuel's energy content is lost to the exhaust in the form of heat and pressure. Some of this can be recovered using a turbocharger, which is effectively a means of extending the compression ratio in the engine's thermodynamic cycle.

Engine Friction. Current engines' fuel economy would improve by around 20% if friction could be eliminated. A high proportion of engine friction occurs when the piston rings slide against the cylinder wall, and in the valve trains. Friction can be reduced by changing the design and materials of rubbing and sliding surfaces or by using improved lubricants.

Pumping Losses. Pumping losses are associated with the passage of the air/fuel mixture through the engine. Eliminating them would improve fuel economy by around 10% in gasoline engines, or rather less in diesel engines. They can be reduced by operating the engine at a lower speed but higher torque[1], by using multiple intake and exhaust valves in each cylinder and by deactivating cylinders at low loads. Tuned intake and exhaust manifolds, and other techniques to reduce pressure drop in the exhaust train, can also contribute.

For the spark ignition engine, the search for improvements tends to focus on air/fuel mixture preparation, combustion chamber design and higher compression ratios. For the diesel engine, emphasis is on improvements in air preparation, fuel injection systems and combustion chamber design.

Engine Management. Much research is also being conducted on improving electronic engine management systems. Car manufacturers consider such technology necessary if vehicles are to meet stricter emission standards while maintaining performance, comfort and fuel efficiency. Electronic engine management systems have consumer appeal because they can also be used for sophisticated control functions in the vehicle. Estimates for the United States show that the fuel economy gain would be limited to less than 1% (OTA, 1991). The gain would be greater in Europe, where electronic fuel injection is not yet the norm. The introduction of lighter vehicle designs and materials will be subject to formidable barriers in the form of R&D costs, production costs, material availability and manufacturing limitations.

Lean-burn Engines. Lean-burn engines provide a potential gain in fuel economy of around 10% although, as mentioned in Chapter 4, their NOx emissions are hard to control (OTA, 1991). Honda and Mitsubishi are marketing lean-burn engines that meet US federal standards but not California standards. Further developments in engines or catalytic converters may make it possible to meet tighter standards with lean-burn engines.

Two-stroke Engines. This technology is being intensively developed by engine manufacturers and may enter the market on a large scale in the next decade. In a two-stroke engine the piston descends and ascends only once for every power stroke. Two-strokes have a higher power output and torque per unit of engine displacement than four-stroke engines, allowing a corresponding downsizing of the engine. They have lower friction and lower heat losses to the engine coolant.

In a two-stroke engine the exhaust gases have to be purged from the cylinder and fresh air/fuel mixture introduced between the power stroke and compression stroke. In conventional two-strokes the charge is introduced via the crankshaft and is effectively pumped into the cylinder by the piston. The fresh charge purges most of the exhaust. Lubricating oil is carried with the charge into the cylinder, and fresh oil has to be added to the fuel to maintain engine lubrication. While this is an elegant piece of engineering, some fuel always passes into the exhaust unburned. Partial combustion of lubricating oil also contributes to exhaust emissions.

---

1.   Turning effort or load.

Two-strokes can be made much cleaner using a more conventional air intake with a large air blower to purge the exhaust and introduce the fresh charge as quickly as possible. Fuel injection is needed, with no fuel being introduced before the exhaust valve is closed. High hydrocarbon emissions are still produced but these can be oxidised catalytically in the exhaust. While further development is needed to reduce emissions, the fuel economy benefit over a conventional two-valve, four-stroke engine could ultimately amount to 12-14% (OTA, 1991).

Transmission. Energy savings of up to 5% may be obtainable by reducing friction in the transmission. There are several possible techniques for obtaining greater savings. An example is the use of a continuously variable transmission managed by a microprocessor linked to the engine management system. Gearbox improvements may ultimately offer from 5% to 15% average energy savings.

5.2   REDUCING TRACTIVE EFFORT REQUIREMENTS

Vehicle Weight. Reducing vehicle weight improves fuel economy. Vehicle weight can be reduced by changing materials, reducing car size and redesigning components to reduce mass. The proportion of steel and cast iron in vehicles is expected to continue to decrease during the next decade, from roughly 60% to 50%. Instead, lighter materials such as plastic, aluminium, nickel and magnesium will be used. The average car in 1997/98 is likely to contain 30% more aluminium than current models. The use of nickel in cars is limited now, but by the mid-1990s is projected to increase by about 40%, mainly alloyed with steel in drive-train components. In practice, vehicle weight may not decrease even if steel is fully displaced, since the weight of ancillary equipment is expected to increase (Martin and Shock, 1989).

A 4-7% fuel economy gain is possible from a 10% reduction in weight: 6% is possible if the engine is downsized to keep car performance constant (OTA, 1991). Reductions in engine size have a multiplier effect — for 1 kg of engine weight removed, 1 kg of structural support can be removed (IEA, 1993). However, engine sizes have not been decreasing recently; improvements in vehicle design capability are being used to provide faster, more comfortable cars.

Aerodynamic Drag. In models redesigned from 1996 to 2001, the average drag coefficient could be reduced by 20%, resulting in an improvement in fuel economy by about 4%. By 2010 an overall reduction in drag coefficients of 35-45% seems obtainable in the United States, with 8-10% fuel economy gains (OTA, 1991).

5.3   THE OUTLOOK FOR CAR FUEL ECONOMY

Technological change is usually gradual, building on existing technology, and therefore to a large extent predictable. Automotive technology is well known, and there is a long lead time involved in technological development, design, tooling and production. Hence it is possible to predict what kind of technology is likely to be produced in the mid- to late 1990s.

There are several approaches to forecasting technical change. They involve looking at: technology already on the market; manufacturers' plans; technology in the research, development and demonstration stages; macroeconomic models incorporating some representation of technical change; and microeconomic models analysing the cost-effectiveness of new technology.

The rate of technical change depends on many factors, such as length of product cycles, the time taken by the industry to redesign models, industry responses to new regulations and consumer responses to changes in vehicle attributes and costs. Each of these factors will affect the ultimate fuel economy achieved (OTA, 1991).

In general, car fuel economy projections are uncertain at best and have tended to be overoptimistic. Inaccurate projections reflect an incomplete understanding of the factors affecting fuel economy. These are complex, and have interactions that are not always evident. The factors include the technologies available and their relative costs and benefits, as well as the many objectives of consumers, of the automotive and oil industries and of politicians and civil servants. To make accurate projections, the analyst would have to be aware of all these factors and their interactions and be able to model them accurately. This is not possible, so analysts have to make compromises in their choice of modelling approaches, depending on their priorities and the type of analysis they are carrying out.

Technical and Economic Potential for Fuel Economy. It is hard to obtain OECD-wide data on the technologies available to improve car fuel economy, their costs and the extent to which they are used. One of several elaborate technical and economic analyses of the evolution of car fuel economy in the United States is briefly discussed below (EEA, 1991a). Such studies significantly improve projections. None have yet been conducted for Europe, but they are very much needed.

Recent studies in the United States indicate new car fuel economy potential for the late 1990s anywhere between 4.8 and 8.2 litres/100 km, depending on the assumptions made (NRC, 1992; OTA, 1991). The bottom of the range could be achieved only through the deployment of a wide range of technology to improve energy efficiency. The upper end of the range is based on a continuation of past trends in the market. The actual fuel consumption could turn out to be even higher if the lack of improvement in the late 1980s continues. The 1991 sales-weighted fuel economy was 8.3 litres/100 km, about the same as the 1986 level.

One indication of technological potential is a "best-in-class" projection[1]. This would result in fuel consumption levels in the United States that are 30% below the current class average for subcompacts, about 25% for compacts and midsize cars and 30% for large cars (NRC, 1992).

A similar study in the United Kingdom (Martin and Shock, 1989), looked at "benchmark fuel consumption" for vehicles in six engine size categories. A fuel economy distribution curve was constructed for the national fleet in each size of vehicle. The lowest 20th percentile was taken as the benchmark for fuel consumption in that engine size range. The UK study indicated benchmark fuel consumption 18-25% lower than the average for each class. This type of analysis can be used as an approximate indication of the current economic potential for fuel economy.

Table 5.1 shows one estimate of the technically achievable fuel economy in the United States, based on a variety of approaches including "best-in-class". It also shows ranges of costs of achieving the fuel economy levels.

---

1.   Based on the assumption that average vehicle fuel economy could reach the level of the current "best-in-class" by a given date. The "best-in-class" fuel economy level is achievable over 10-15 years, assuming that by that time most existing high-volume car production lines will have been retooled.

Table 5.1
Technically Achievable Fuel Economy for New Vehicles, United States, 2006

Vehicle Size Class	Ranges of Technically Achievable Fuel Economy in 2006 Model Year[1]	Retail Price Increase for Improved Fuel Economy in 2006 Model Year[2]
	(litres/100 km)	(1990 US$)
Passenger Cars		
Subcompact	6.07 to 5.88	500-1 250 1 000-2 500
Compact	6.95 to 6.23	500-1 250 1 000-2 500
Midsize	7.39 to 6.76	500-1 250 1 000-2 500
Large	7.87 to 7.17	500-1 250 1 000-2 500
Light Trucks		
Small pickup	8.16 to 7.39	500-1 000 1 000-2 000
Small van	8.45 to 7.89	500-1 250 1 000-2 500
Small utility	9.10 to 8.16	500-1 250 1 000-2 500
Large pickup	10.29 to 9.46	750-1 750 1 500-2 750

1. Compliance with Tier I emission standards of the 1990 US Clean Air Act Amendments is assumed, as is performance equivalent to that of 1990 vehicles. The estimates are based on the introduction of existing near-commercial technology, taking account of past trends in fuel economy and the "best-in-class" fuel economy of the existing fleet.

2. Estimated increment in first cost to consumers of improved fuel economy. Changes in vehicle emission and safety standards will add to these costs.

Source: NRC (1992).

Fuel economy increases will be associated with a large number of individual changes in car technology. Estimates of the benefits of some changes are presented in **Table 5.2**. The corresponding cost estimates are presented in **Table 5.3**. The estimates must be interpreted relative to a fixed technological baseline, and are calculated assuming constant car performance and interior volume. The figures in this table cannot simply be added up to a sum total of future fuel economy.

Market Potential for Fuel Economy. Table 5.4 shows national estimates and projections for average new passenger car fuel consumption submitted to the IEA by some OECD Member countries. Based on this information, as well as on extrapolation from the most recent data for new car registrations for OECD countries (IRF, 1990), average new car fuel economy for 2005 could reach 6.8 litres/100 km in North America and 6.9 litres/100 km in OECD Europe. Recent trends do not bear out these projections.

Table 5.2
Effect of Technical Changes on Fuel Economy

	BASELINE	EEA Estimate	Manufacturers' Estimates
		% Fuel Saving	% Fuel Saving
ENGINE TECHNOLOGY			
General			
Roller cam followers	Flat followers	2.0	0.8-3
Friction reduction by 10%	Base 1987	2.0	0.5-2
Accessory improvement	Conventional	0.5	0 -1.4
Fuel systems			
Throttle-body fuel injection	Carburettor	3.0	0.8-3.4
Multipoint fuel injection	Carburettor	5.0	2.5-6.0
Valve Train			
Overhead camshaft	Overhead valve	3.0	0.8-3.5
4 valves per cylinder	2 valves	5.0	2.0-4.5
Variable valve timing	Fixed timing	6.0	1.5-3.0
Reduced Number of Cylinders			
4-cylinder	6-cylinder	3.0	-3.0-0
6-cylinder	9-cylinder	3.0	0
TRANSMISSION TECHNOLOGY			
Torque converter lock-up	Open converter	3.0	2.0-3.2
Electric transmission control	Hydraulic	0.5	0 -0.6
4-speed automatic	3-speed automatic	4.5	1.8-4.0
5-speed automatic	3-speed automatic	7.0	3.5-5.0
Continuously variable transmission	3-speed automatic	8.0	3.0-5.5
ROLLING RESISTANCE, AERODYNAMICS AND WEIGHT			
Front-wheel drive	Rear-wheel drive	10.0	0 -3.0
Aerodynamics	Base	2.3	1.2-3.1
Weight reduction by 10%	Base	6.6	5.0-8.0
Electric power steering	Conventional	1.0	0.5-1.5
Advanced tyres by 10%	Base	1.0	0.5-1.0
Advanced lubricants	Conventional	0.5	0.2-0.5

Source: EEA (1991b).

Table 5.3
Costs of Technical Improvements

| | | EEA Cost Estimate (1988 $) Engine Size (Number of Cylinders) | | |
	BASELINE	4	6	8
ENGINE TECHNOLOGY				
General				
Roller cam followers	Flat followers	16	24	32
Friction reduction by 10%	Base 1987	30	40	50
Accessory improvement	Conventional	12	12	12
Fuel Systems				
Throttle-body fuel injection	Carburettor	42	70	70
Multipoint fuel injection	Carburettor	90	134	150
Valve Train				
Overhead camshaft	Overhead valve	110	160	200
4 valves per cylinder	2 valves	140	180	225
Variable valve timing	Fixed timing	140	200	267
Reduced Number of Cylinders[a]				
4-cylinder	6-cylinder	0	(300)	(550)
6-cylinder	8-cylinder	300	0	(250)
TRANSMISSION TECHNOLOGY				
Torque converter lock-up	Open converter	50	50	50
Electric transmission control	Hydraulic	24	24	24
4-speed automatic	3-speed automatic	225	225	225
5-speed automatic	3-speed automatic	325	325	325
Continuously variable transmission	3-speed automatic	325	325	325
ROLLING RESISTANCE, AERODYNAMICS AND WEIGHT				
Front-wheel drive	Rear-wheel drive	240	240	240
Aerodynamics	Base	40	40	40
Weight reduction by 10%	Base		— varies —	
Electric power steering[c]	Conventional	45	45	45
Advanced tyres by 10%	Base	18	18	18
Advanced lubricants	Conventional	2	3	3

Sources: EEA (1991b), SRI (1991).

Table 5.4

National Estimates and Projections of New Passenger Car Fuel Consumption

(litres/100 km)

	ACTUAL							Total decrease for period considered (%)	PROJECTIONS			
	1979	1983	1984	1985	1986	1987	1988		1990	1995	2000	2010
OECD NORTH AMERICA												
Canada	11.4	8.5	8.5	8.5	8.4	8.3	8.1	-28.9	8.1	7.9	7.7	7.6
United States	11.6	8.9	8.7	8.5	8.4	8.3	8.2	-29.3	8.3	7.6	7.1	6.3
OECD EUROPE												
Western Germany	9.6	8.1	7.8	7.6	7.5	7.7	7.9	-17.7	7.9	7.7	7.4	6.8
Italy	8.3	7.3	7.0	7.0	6.8	6.8	6.8	-18.1	6.8	7.1	7.1	n.a.
Netherlands[1]	n.a.	7.5	n.a.	7.2	n.a.	7.2	n.a.	n.a.	7.3.	7.0	6.3	5.5
Spain	8.7 (1980)	n.a.	n.a.	7.4	n.a.	n.a.	7.4	-14.9	7.2	7.0	6.7	6.4
Sweden[2]	9.2	8.6	8.6	8.5	8.4	8.3	8.2	-10.9	n.a.	n.a.	n.a.	n.a.
United Kingdom[3]	9.0	7.9	7.6	7.5	7.5	7.4	7.4	-17.8	n.a.	n.a.	n.a.	n.a.
OECD PACIFIC												
Australia[4]	11.2	9.5	9.5	9.5	9.3	9.4	9.1	-18.8	n.a.	8.7	8.2	n.a.
Japan	8.6	7.8	7.8	8.0	8.3	8.6	8.6	0	n.a.	n.a.	n.a.	n.a.
New Zealand[5]	10.5	9.7	9.4	9.2	9.0	9.0	9.0	-14.3	8.7	6.5	7.8	7.6 (2005)

1. 1989: 7.25 litres/100 km; the data for 1983-1989 are weighted (1/3-1/3-1/3), based upon the ECE cycle, for 80% of new passenger cars. The projections for 1990-2010 are not agreed government policy but estimates derived from the policy goal of reducing energy consumption in cars by 35-40% between 1986 and 2010 (20% between 1990 and 2000).
2. 1989: 8.4 litres/100 km.
3. Figures obtained using a weighted average of new car models registered, excluding diesel and four-wheel-drive vehicles.
4. Fuel consumption tests were conducted in accordance with AS 2077 before 1986 and in accordance with AS 2877 thereafter. The change "understates" fuel consumption by 2.75% in comparison with earlier data series. With the new test, fuel consumption was 9.56 litres/100 km in 1986, 9.66 litres/100 km in 1987 and 9.35 litres/100 km in 1988. Figures for 1995 and 2000 are industry projections.
5. 1989 = 8.9 litres/100 km; the figures for 1989 and 1990 are actual. The reduction from 1979 to 1990 is 17%. All these data are weighted average figures for measurements across different capacity engines.

n.a. = Not available.

Note: Test procedures differ between countries, so direct comparisons should not be made between national fuel consumption figures.

Sources: IEA (1991b, 1991c); *Energy Policies and Programmes of IEA Countries* (Paris, OECD, various years); and country submissions for IEA publications.

Tested Fuel Economy vs. On-Road Fuel Economy

Official new car fuel economy data do not necessarily reflect the average fuel economy of cars on the roads. According to test figures, new car fuel economy has improved by roughly 30% in Germany, France and Japan since 1973, but this is not reflected in reduced fuel use. In the United States studies indicate that cars consume 20% more fuel on the roads than in official tests.

Many factors are cited for discrepancies between on-road and tested fuel consumption. Tests may not provide an accurate reflection of real driving patterns. In most countries cars are not tested with auxiliaries such as air conditioning in operation. Some official tests do not include cold starts, which can result in discrepancies as high as 30%. Levels of efficiency are not directly comparable between countries because of differences in national test cycles. The variation between countries in official test fuel economy data, arising from such differences, is estimated at 10-20%.

Figure 5.2
Car Energy Use per Vehicle-km, 1970 - 1990
(MJ/Vehicle-km)

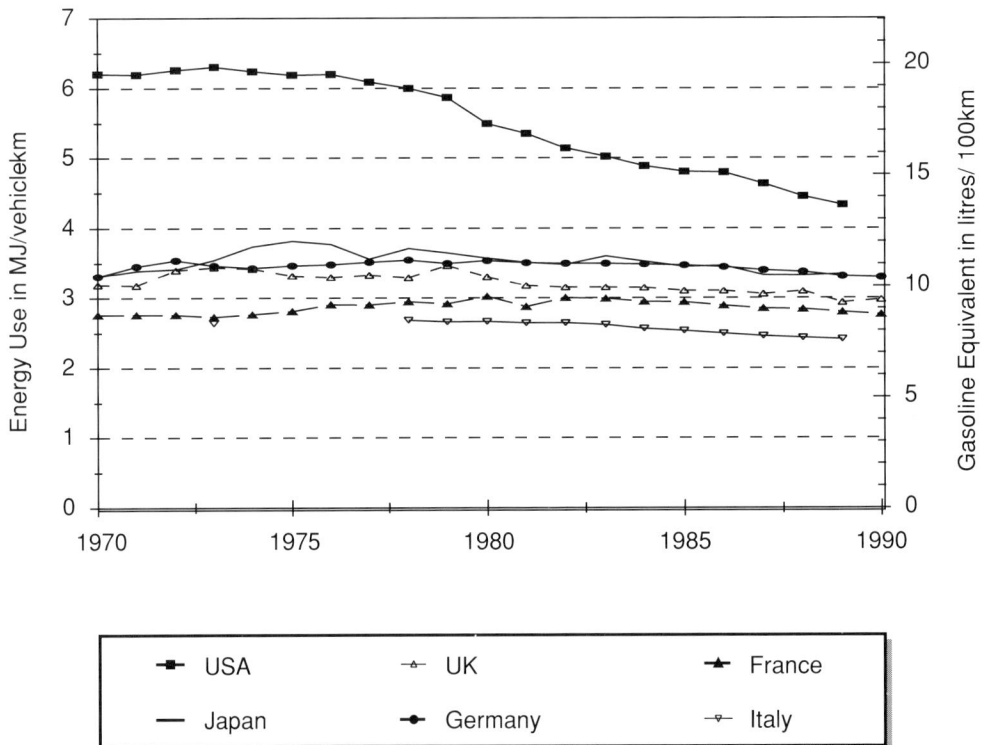

Source: Schipper (1992).

Note: Energy use is in lower heating values.

Including the light truck market in the US estimate could lead to 10-20% higher projected fuel consumption. In the United States light trucks account for roughly one-third of light-duty vehicle sales and their fuel economy is roughly 20% poorer than that of average passenger cars (NRC, 1992).

National estimates are not a good basis for international comparisons, due to a lack of standardisation in methodology (see **Box**). Tested new car fuel consumption is also unlikely to give an accurate reflection of on-road fuel economy. **Figure 5.2** shows the actual progress in on-road fuel economy in several countries over the period 1970 to 1990.

Table 5.5 shows the results from a business-as-usual scenario in a study of fuel economy in cars in the United States. In this scenario the deterioration in fuel economy of the late 1980s is reversed, resulting in a 13.5% reduction in fuel consumption in the period 1990-2001. Assuming national on-road fuel economy improves at the same rate, a fuel consumption level of 11 to 11.5 litres/100 km might be expected for the United States as a whole in the year 2001.

Table 5.5
A US Scenario of New Car Fuel Economy
(litres/100 km)

	1988	1990	1995	2001
Domestic Vehicle	8.73	8.66	8.45	7.46
Imported Vehicle	7.53	7.97	7.68	6.94
New Vehicle Fleet	8.27	8.39	8.13	7.26

Source: NRC (1992).

Regional Differences. The automotive industry is becoming increasingly globalised and the application of new technology is becoming more uniform. Despite this, average fuel economy will not evolve in parallel in North America and OECD Europe, partly because of different consumer preferences for vehicle size and performance criteria.

The European fleet has a high proportion of small cars, so there is less room for downsizing the fleet than in North America. If anything, there is a tendency to move "up-market", towards larger and/or more powerful cars. In addition, demand for air conditioning and automatic transmissions will probably increase in Europe.

Conversely, the North American market is dominated by large cars, leaving greater scope for downsizing. This has been a major factor in the more rapid improvement in average new car fuel economy in the United States than in Europe over the last two decades. The trend is expected to continue in the short to medium term.

CHAPTER 6

LIFE-CYCLE ANALYSIS OF ALTERNATIVE FUELS

Interest in alternative transport fuels in the OECD developed in several phases. In response to fuel scarcity during the two world wars, Germany developed several ways to convert coal to liquid fuels. Other countries developed ways of using gaseous fuels. The second phase of development occurred in the 1970s and 1980s, when several of these fuel technologies were revived in response to the oil crises and concern about security of supply. The current phase of alternative fuel development is mainly a response to environmental concerns. Technology originally developed by Germany is still used in commercial plants in South Africa, and form the basis of many recent alternative fuel initiatives.

A large amount of development work is now focused on zero-emission vehicles for urban areas. The automotive industry is working especially hard on batteries for electric vehicles and on vehicle designs that would make the use of battery power feasible. There is also considerable interest in the use of CNG, LPG, methanol, ethanol and hydrogen. All these alternative fuels could reduce urban pollution. They may also have the potential to reduce greenhouse gas emissions.

Fuel substitution can reduce greenhouse gas emissions in a number of ways. Alternative fuels may contain less carbon than conventional fuels, resulting in lower tailpipe emissions of CO_2. They may involve less energy use for production and processing, resulting in lower upstream CO_2 emissions. In either case, they may also lead to a reduction in emissions of greenhouse gases other than CO_2.

This chapter looks at some of the basic issues surrounding the use of alternative transport fuels: technical viability, environmental acceptability and life-cycle greenhouse gas emissions. The viability of all transport fuels depends on a wide range of issues, including safety, fuel economy, costs and engine performance. The chapter touches on most of these issues. The availability in the OECD of various fuels is briefly discussed, and their technical features are outlined. The fuels analysed are gasoline, reformulated gasoline, diesel, natural gas (compressed or liquefied), LPG, methanol from natural gas or wood, ethanol from maize or wood, electricity — with any power generation mix — and hydrogen in internal combustion engines. **Chapter 7** makes a detailed examination of the costs of using some alternative fuels.

6.1   ALTERNATIVE FUEL TECHNOLOGY

Alternative-fuel vehicles are used in many countries, both within and outside the OECD. In OECD countries, for light-duty applications, these are generally converted gasoline vehicles that can use both gasoline and an alternative. Dual-fuel vehicles are a way to introduce alternative transport fuels but may not be the best technical solution in the long term. Generally, dual-fuel vehicles have suboptimal performance because the engine design is optimised for the conventional fuel.

When fully developed, vehicles purpose-built for alternative fuels could potentially perform as well as, or better than, conventional gasoline-powered vehicles. They might also have lower emissions. Many of the fuels require considerable development before they can achieve this potential.

Table 6.1 indicates the development status of some of the alternative fuels and electric vehicles discussed in this study. All fuels except hydrogen have been successfully demonstrated in on-road vehicle fleets. Diesel vehicles are fully commercial, and purpose-built CNG vehicles are likely to be commercialised during the 1990s. Unblended alcohol fuels have been used commercially outside the OECD in specially manufactured vehicles. Internal combustion engines and fuel cells using hydrogen have been slow to develop and must be classified as long-term options. The development of liquid alternative fuels has progressed the farthest, since they are most compatible with existing technology.

Table 6.1
Status of Alternative Fuel Technology for Cars

	Technical Development	Commercialisation
Diesel (indirect-injection engines)	Mature, continuing development.	Widespread, growing market.
Diesel (direct-injection engines)	Under development.	Small but growing market.
CNG (conversions)	Mature.	Niche markets where actively promoted.
CNG (from manufacturer)	Demonstrated.	Production lines under construction.
LPG (conversions)	Mature.	Niche markets especially where actively promoted.
LPG (from manufacturer)	No significant developments.	
Ethanol (up to E15)	Mature — used in gasoline vehicles.	Widespread use where promoted.
Ethanol (E85 to E100)	Development required for OECD use.	Widespread use in some non-OECD countries.
Methanol (up to M15)	Mature — used in gasoline vehicles.	Some use where promoted.
Methanol (up to M85)	Prototype fleets.	Some production vehicles available.
Hydrogen	Considerable technical difficulties.	Distant.
Electric Vehicles	Considerable development needed to match gasoline vehicle performance.	Some use. May achieve considerable market share because of anti-pollution legislation in some countries.

6.2   THE LIFE-CYCLE EMISSIONS MODEL

To understand the effects of fuel switching on greenhouse emissions, it is necessary to look at the whole fuel and vehicle life-cycle. Significant emissions can occur in the extraction of raw materials, their conversion to final fuels and the transport of raw materials and fuels. As fuel switching affects vehicle design, the emissions associated with vehicle manufacture can also be affected.

Upstream emissions depend on a number of factors. Where electricity is a significant input, either for fuel production or for battery charging, the power generation mix is important. Similarly, refinery energy use depends on the type of crude oil being used, the product mix and the equipment installed.

In this study, life-cycle emissions have been calculated using a spreadsheet-based emissions model[1]. This programme calculates upstream emissions associated with fuel use by vehicles, as well as emissions from vehicle operation and manufacture. The model's main output is the global warming effect of the emissions per kilometre of vehicle travel. **Annex 5** contains a detailed description of the model.

The life-cycle greenhouse gas emission analysis is performed for gasoline, diesel, other fuels and electricity in light-duty vehicles. Dual-fuel or hybrid vehicles are excluded.

The model is used to calculate emissions of CO_2, carbon monoxide, methane, NO_x, N_2O and VOCs. It uses each gas's GWP to convert emissions to CO_2 equivalents. The model can run with different sets of GWPs; the values used here are given in **Table 2.1**.

The model is based upon a detailed analysis of every process involved in the production, distribution and use of each fuel. Designed for analysis of alternative road transport fuels in the United States, it was originally configured around a US energy supply and demand structure for 2000. The IEA has modified the model inputs to represent the car/fuel supply chains in two OECD regions: North America and Europe. The main changes are in the power generation mix, refinery sector energy use and emissions, and vehicle emissions. The power generation mix is based on the IEA World Energy Outlook (IEA, 1991b).

6.2.1   Stages in the Life-cycle Analysis

The emissions model accounts for emissions from: the tailpipe of the vehicle; the manufacturing and assembly of materials to produce the vehicle; fuel production and distribution, including refuelling; and fuel feedstock recovery. The emissions from all stages are spread over the total distance travelled by the car during its active life, assumed to be 110 000 miles (177 000 km).

Vehicle disposal or recycling is not considered here. If the organic material in the vehicle is burned at the end of its life, the life-cycle emissions rise by about 1-2%. If landfill disposal is used, methane emissions can result, with global warming impact amounting to up to 5% of life-

---

1.   Developed by M. DeLuchi at the University of California, Davis (DeLuchi, 1992).

cycle emissions. The use of more easily recyclable materials has the potential to reduce these emissions, as well as energy use in vehicle manufacturing. This could give a reduction of the order of 5% in life-cycle greenhouse gas emissions, but further study is required to quantify the effect for individual OECD countries.

Vehicle Manufacturing. Greenhouse gas emissions from vehicle manufacturing for all vehicles are based on the material content of a base gasoline vehicle as well as on the energy mix in the manufacturing region. Differences in engine materials and fuel storage systems from one fuel and vehicle type to another (including batteries for electric vehicles) will lead to variations in manufacturing emission levels.

Fuel Supply. All emission factors in the analysis are based on US data. Emission factors from the US Environmental Protection Agency (EPA) database are used for each type of energy use. European measurements of emissions from many of the processes modelled are now becoming available, but could not be included in this study. The model divides fuel supply into four stages:

- Greenhouse gas emissions first arise in **feedstock recovery**, principally from fuel use in the recovery process. They also include emissions from gas flaring in oil fields, methane emissions from coal mines and leaks of methane and CO_2 during natural gas extraction.

- Emissions arise during **fuel and feedstock transport** from energy consumption by crude oil tankers, trains, trucks and pipeline compressors.

- **Fuel production** creates greenhouse gas emissions, partly from process energy consumption and partly from leaks (particularly of VOCs and methane). Emissions considered here come from oil refineries, alcohol fermentation facilities, synthesis gas plants, natural gas processing plants, hydrogen production and power generation. In processes such as oil refining, which have several products, emissions are allocated to products based on estimates of the process energy needed for each product. **Table 6.2** shows the energy inputs assumed for each fuel conversion process in the model.

- **Fuel distribution** may involve the use of trucks, trains, ships and pipelines. Greenhouse gas emissions differ depending upon the mode of transport used. The emissions model accounts not only for emissions from vehicle operation in each transport mode, but also for those arising in fuel supply for transport. The model includes the average distance over which fuels must be transported by each mode in the relevant OECD region as well as emissions from building and maintaining transport facilities.

Vehicle Tailpipe. Tailpipe emissions of CO_2 in the model reflect a complete accounting of fuel carbon. CO_2 emission factors are a function of vehicle energy consumption (in MJ/km) and of the fuel carbon content (in grams/MJ). The CO_2 emission factors exclude carbon that is not oxidised to CO_2 but emitted as methane, carbon monoxide and VOCs.

For non-CO_2 greenhouse gases, emissions per kilometre are entered directly into the model for each type of vehicle. The emission factors for conventional pollutants (NO_x, carbon monoxide and VOCs) reflect the standards currently planned for each OECD region. It is assumed that all light-duty vehicles, regardless of the fuel they use, meet these limits. North American tailpipe emission limits are assumed to be those defined for transitional low-emission vehicles in California (discussed in **Chapter 9**). Tailpipe emissions of greenhouse gases in Europe are

Table 6.2

Assumed Energy Inputs in Fuel Supply
(kJ of energy use/MJ of final fuel)

Fuel	Chemicals Used in Agriculture	Raw Material Extraction, Harvesting, etc.	Raw Material Transport	Conversion to Final Fuel	Fuel Distribution	Compression or Liquefaction
Unleaded Gasoline from Crude oil		27	12	146	8	
Reformulated Gasoline from Crude oil		25	12	185	8	
Low-Sulphur Diesel from Crude oil		29	13	70	9	
CNG		28		25	36	50
Methanol from Natural Gas		79	22	1 540*	38	
Methanol from Wood	69	62	18	1 770*	24	
Ethanol from Maize	194	100	26	580	28	
Ethanol from Wood	92	82	24	2 350*	18	
Hydrogen				Depends on source of hydrogen	100	310

* Includes energy in feedstock.

assumed to be 5% higher than in North America. Vehicles are assumed to meet the limits in both OECD regions using advanced catalytic converter technology (i.e. heated catalysts), exhaust gas recirculation and electronic engine management. N_2O and methane emissions are unregulated, so the best available experimental data have been used to reflect actual emissions.

The car used as the base for analysis in this study is based on the Volkswagen Golf Boston II. It is assumed to have fuel consumption in 2005 of 7.3 litres/100 km. This is a typical European car. It is used in the model for both North America and OECD Europe to draw out any differences associated with fuel supply systems. If current trends continue, cars in North America are likely to be larger and have higher fuel consumption than this, as discussed in **Chapter 5**.

The energy use of alternative vehicles is calculated in the model from that of the base car, taking account of drive-train efficiency and vehicle weight. Drive-train efficiency is exogenous in each case, but weight changes are calculated within the model — for example, it derives the weight of fuel tanks from the vehicle range. All alternative vehicles except electric vehicles are assumed to have the same range as gasoline vehicles. Energy use in vehicle operation is assumed to increase by 6% for each 10% increase in vehicle weight.

Electric vehicles require an independent analysis, as they cannot substitute directly for gasoline vehicles. The market for electric vehicles is limited by their poor performance and range. They are likely to be used mainly in urban areas where they are required by legislation. For this reason, a supplementary comparison is made between gasoline and electric vehicles, based solely on an urban driving cycle. Additional calculations have been made to show the effect of the fuel mix for power generation on life-cycle greenhouse gas emissions of electric vehicles.

6.3 ALTERNATIVE FUELS AND THEIR LIFE-CYCLE EMISSIONS

6.3.1 Gasoline

Gasoline is a product of the processing of crude oil. A variety of processes is used to convert distillation products to gasoline components. Gasoline accounts for about 22% of oil product demand by volume in OECD Europe and 42% in the United States.

For gasoline blends, the octane number is one of the most important indicators of fuel quality. While the basic fuel quality affects the octane number, it can also be raised by the use of components such as tetra-ethyl lead. The use of lead is being phased out in all regions of the OECD and the process is virtually complete in North America and Japan. The phase-out requires changes to refinery processes and/or the addition of high octane components, such as methyl tertiary butyl ether (MTBE).

Reformulated Gasoline. Both gasoline engines and fuel quality have evolved in response to market conditions. Environmental constraints are currently a major driving force behind this process. Since the adoption of the US Clean Air Act Amendments, reformulated gasoline has been introduced in ozone non-attainment areas[1]. The progressive tightening of vehicle emission

---

1.   Areas where the fourth highest recorded concentration of ozone in any 24-hour period during the past three years exceeded the federal standard for ozone.

standards in the United States is forcing the development of low-emission vehicle technology to its limit. Further pollution abatement through vehicle technology will be increasingly expensive. Some pollution reduction can be achieved through changes in fuel standards, which may in some cases be more cost-effective than changes in vehicle engineering.

Life-cycle Emissions. Reformulating gasoline to US Clean Air Act standards produces virtually the same life-cycle greenhouse gas emissions, in grams of CO_2 equivalent per kilometre, as the use of premium unleaded gasoline. Energy consumption and CO_2 emissions are higher, but are offset by lower non-CO_2 greenhouse emissions.

The majority of life-cycle CO_2 equivalent emissions for gasoline and reformulated gasoline cars are emitted in vehicle operation (72% in Europe, 73% in North America). The production and transport of fuel and feedstocks account for 17-18%. **Figure 6.1** shows the breakdown by life-cycle stage and by greenhouse gas.

Figure 6.1
Gasoline Car Life-cycle Greenhouse Gas Emissions

Based on North American vehicles and energy supply system. Life-cycle emissions for
new cars in OECD Europe in 2000 are not expected to be significantly different.

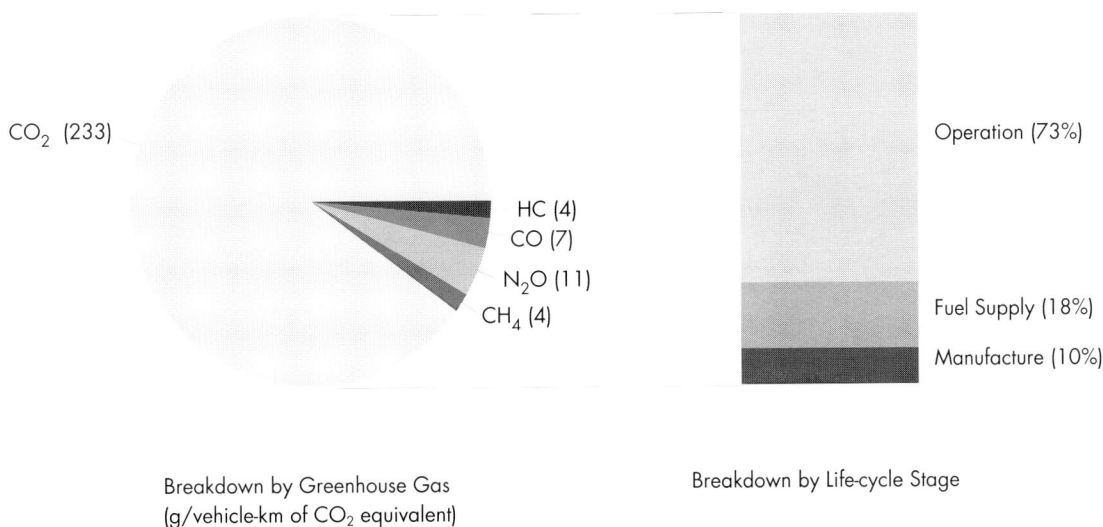

CO_2 (233)

HC (4)
CO (7)
N_2O (11)
CH_4 (4)

Operation (73%)

Fuel Supply (18%)

Manufacture (10%)

Breakdown by Greenhouse Gas
(g/vehicle-km of CO_2 equivalent)

Breakdown by Life-cycle Stage

6.3.2   Diesel

Diesel fuel comprises middle distillate "straight-run components" from crude oil (boiling between 150°C and 400°C), and streams from secondary processes such as hydrodesulphurisation and cracking. Diesel fuel properties are highly dependent on the quality of the crude oil feedstock.

Life-cycle Emissions. Tailpipe emissions from diesel vehicles account for a larger share of life-cycle greenhouse gas emissions than those from gasoline vehicles. Fuel production and distribution produce a smaller share because producing diesel is generally less energy intensive than producing gasoline.

The composition of greenhouse gas emissions from consumption of diesel in cars is dominated by CO_2. The level of these greenhouse gas emissions (in CO_2 equivalent grams per kilometre) is about 20% below that for gasoline (see **Figure 6.2**).

Figure 6.2
Diesel Car Life-cycle Greenhouse Gas Emissions

Based on North American vehicles and energy supply system. Life-cycle emissions for new cars in OECD Europe in 2000 are not expected to be significantly different.

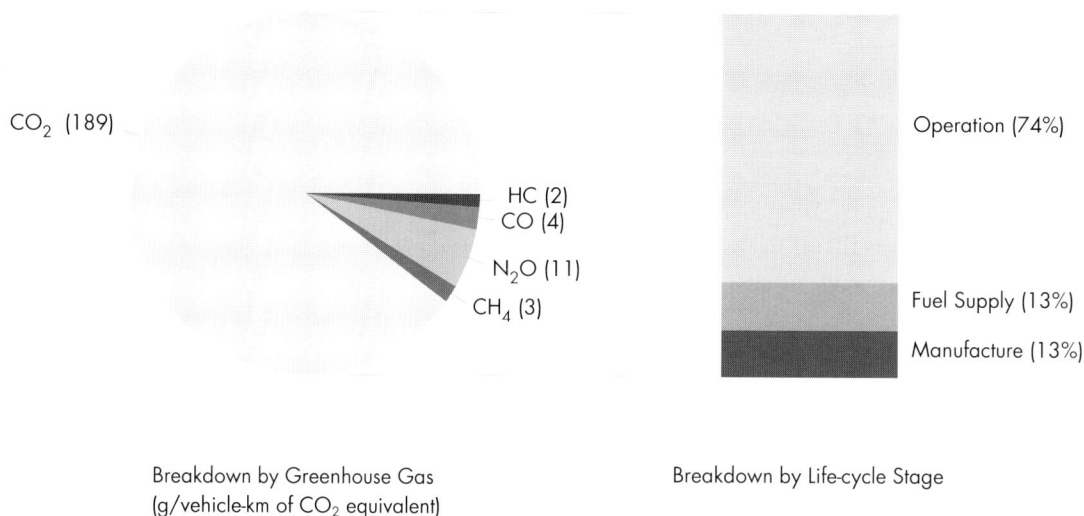

CO_2 (189)

HC (2)
CO (4)
N_2O (11)
CH_4 (3)

Operation (74%)

Fuel Supply (13%)
Manufacture (13%)

Breakdown by Greenhouse Gas
(g/vehicle-km of CO_2 equivalent)

Breakdown by Life-cycle Stage

Tighter constraints are being introduced on the quality of diesel fuel in both OECD Europe and North America. This will oblige refiners to use more energy-intensive processes to reduce the amount of sulphur in the fuel. The rising share of diesel demand in Europe also has implications for refinery energy use. A greater proportion of diesel will be produced by cracking heavier fractions of crude oil, increasing refinery energy use. Constraints on diesel tailpipe emissions, especially for particulates, could result in a fuel economy loss. All of these factors will raise the life-cycle emissions of CO_2 associated with diesel use.

6.3.3 Liquefied Petroleum Gas (LPG)

LPG is produced mainly in the production and refining of oil and natural gas. The transport sector only represents about 5% of total OECD LPG consumption and there is scope to increase its use.

Transport consumption of LPG is currently limited to areas in which there are good supplies (e.g. the Netherlands) or in which its clean burning characteristics are particularly valued (especially in urban applications such as taxis).

LPG consists mainly of propane and butane, which liquefy at moderate pressure and ambient temperature and thus are convenient to store. The proportion of propane to butane in automotive LPG varies by region and is generally controlled by national specifications.

Storage and Handling. LPG has a higher energy content per unit of mass but lower volumetric energy content than gasoline. For the same vehicle range a larger, heavier fuel storage system must be used, resulting in a slight fuel economy penalty.

Engine. LPG, whether burnt in dedicated or dual-fuelled gasoline/LPG engines, produces relatively low emissions because, being a gas at ambient temperature and pressure, it mixes readily with air and does not require cold start enrichment. LPG engines have much lower carbon monoxide emissions than gasoline engines and consequently are used in indoor applications. LPG's octane rating is roughly 5-10% higher than that of gasoline. This allows higher compression ratios to be used, so engines optimised for LPG can be more efficient than gasoline engines. Even in converted gasoline engines, eliminating the need for cold start enrichment contributes to fuel economy.

Life-cycle Emissions. LPG has a lower carbon content per unit of energy than gasoline or diesel and its refinery processing uses very little energy. Per kilometre, the energy needed to liquefy the fuel is less than that to fill a CNG cylinder. Evaporative emissions or leaks in fuel distribution and use are much lower than for gasoline, and tailpipe emissions of non-CO_2 greenhouse gases are also lower.

Life-cycle greenhouse gas emissions from the LPG vehicle are shown in **Figure 6.3**. The overall impact is about 24% lower than that of the gasoline vehicle.

Figure 6.3
LPG Car Life-cycle Greenhouse Gas Emissions

Based on North American vehicles and energy supply system. Life-cycle emissions for
new cars in OECD Europe in 2000 are not expected to be significantly different.

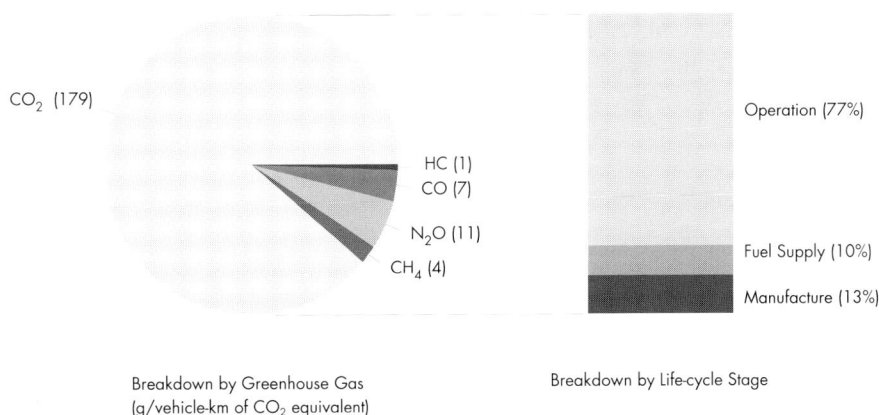

CO_2 (179)

HC (1)
CO (7)
N_2O (11)
CH_4 (4)

Operation (77%)

Fuel Supply (10%)

Manufacture (13%)

Breakdown by Greenhouse Gas
(g/vehicle-km of CO_2 equivalent)

Breakdown by Life-cycle Stage

6.3.4 Natural Gas

Demand for natural gas in the OECD is expected to rise rapidly over the coming decade, driven largely by increasing demand for power generation (IEA, 1992a). This will have a significant impact on the price outlook and availability of the fuel by the turn of the century. Reserves are currently comparable with those of petroleum. Less effort has been devoted in the past to gas exploration than to oil exploration, so considerable reserves may remain to be discovered.

Storage and Handling. Natural gas can be transported in either compressed (CNG) or liquid form (LNG). Although some vehicle tests have been done using LNG, the development of CNG vehicles is far more advanced, with numerous commercial programmes in OECD countries. LNG vehicles are not considered in detail in this report.

A CNG vehicle requires large, heavy gas cylinders to have a reasonable operating range. The added weight leads to some reduction in top speed, acceleration capability and fuel economy. In small cars, the cylinders can occupy most of the cargo space.

CNG storage systems have five times the volume and weight of gasoline tanks containing the same energy. A typical fuel tank containing 75 litres of gasoline weighs 55 kg, whereas a steel cylinder holding CNG with the same energy content can weigh around 270 kg. Because of the cylinder weight and cost, the storage capacity of a CNG vehicle must be carefully matched to its daily operating range.

Aluminum or plastic composite cylinders for CNG are lighter than steel, but more expensive. There is considerable interest from the aerospace industry in the possibility of using carbon fibre composite production facilities to make high-pressure cylinders. The facilities were installed to produce missile cases, for which demand has declined significantly. Carbon fibre composite cylinders could therefore become available at relatively low cost. They would weigh roughly 50% less than steel cylinders.

Advanced gas storage systems are possible, using activated carbon, layered clay or other material to adsorb large volumes of gas at low pressure. The advantage of such systems would be low storage pressure (about 35 bar) and net weight reductions despite the additional weight of the adsorbent material.

Concern about safety could act as a barrier to public acceptance of CNG as a vehicle fuel. Since it is lighter than air, however, it cannot pool in enclosed areas and lead to suffocation or risks of explosion. High-pressure storage does enhance the potential for explosive discharge, and international CNG cylinder safety standards are needed.

Engine. Most of the light- and heavy-duty CNG vehicles in operation in the world are gasoline vehicles converted to dual-fuel capability. In gasoline engines converted to run on CNG, the gas is added to intake air through a variable valve in the inlet manifold or carburettor. Gas in the engine intake displaces air so that less fuel can be burned, causing a fall in maximum power output of 15-20% in a typical gasoline- or diesel-optimised engine (Stephenson, 1991). If the engine were optimised for CNG, the fall in full throttle output would be about 5-10%. Acceleration-rate losses of up to 20% are common in dual-fuel conversions. The loss of power can be compensated by the use of a larger engine or a turbocharger, adding to the cost (USDOE, 1990a).

In a CNG-diesel engine, CNG is added to intake air as in converted gasoline engines. To ignite the air-fuel mixture, it is necessary either to use spark-plugs or to inject a small amount of diesel fuel into the compressed mixture in the combustion chamber. Dual-fuel diesel engines are about 20% more energy efficient than spark ignition engines.

Natural gas has an octane rating about 30-40% higher than that of gasoline, making it appropriate for use in an Otto-cycle engine with a very high compression ratio. The gaseous state of the fuel improves throttle response, minimises spark-plug fouling, reduces oil contamination (and consequently engine wear and maintenance) and makes cold starting easier. Natural gas also has wider flammability limits than gasoline, allowing it to operate in leaner conditions with greater stability (Heath, 1991). A lean air/fuel mixture for natural gas can improve fuel efficiency by 5-10%, although NO_x emissions are then hard to control (Cannon, 1989).

An engine specifically designed to take advantage of the properties of CNG would have better fuel efficiency and power at full load than adapted engines. Emission levels from a purpose-built CNG vehicle are expected to be lower than those from an equivalent gasoline vehicle, except for NO_x and methane. Fuel combustion is expected to be more complete than for gasoline and to give rise to fewer cold-start problems and emissions.

Several technical problems remain to be solved in the natural gas engine, including frequent leaks, problems with the gas regulators and difficulties with high engine temperatures (Cannon, 1989). Special catalysts need to be developed to reduce emissions of NO_x and methane from CNG vehicles.

Life-cycle Emissions. Figure 6.4 shows the breakdown of life-cycle emissions expected from a CNG car. Tailpipe CO_2 emissions are lower than those of a gasoline vehicle. The difference reflects higher fuel efficiency and a lower emission rate of CO_2 per megajoule of natural gas.

Figure 6.4
CNG Car Life-cycle Greenhouse Gas Emissions

Based on North American vehicles and energy supply system. Life-cycle emissions for
new cars in OECD Europe in 2000 are not expected to be significantly different.

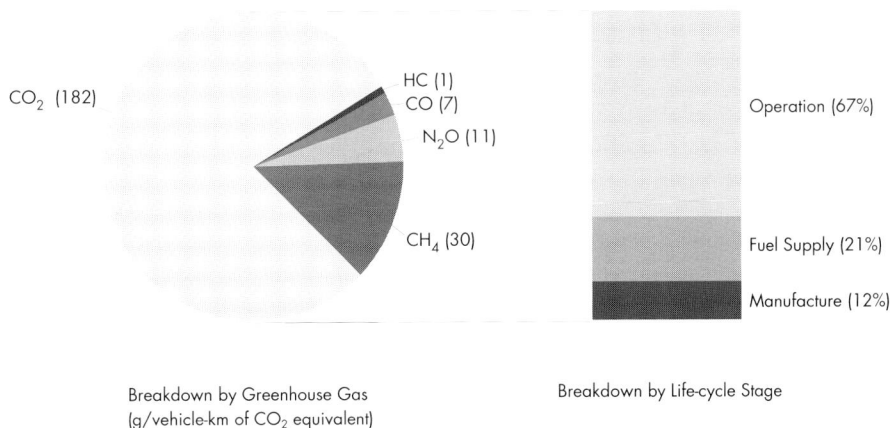

CO_2 (182)

HC (1)
CO (7)
N_2O (11)

CH_4 (30)

Operation (67%)

Fuel Supply (21%)

Manufacture (12%)

Breakdown by Greenhouse Gas
(g/vehicle-km of CO_2 equivalent)

Breakdown by Life-cycle Stage

In terms of their greenhouse impact, methane emissions play a larger role in the natural gas fuel cycle than in other fuel cycles. This reflects the effect of gas leaks throughout the cycle, including higher methane emissions at the tailpipe. Several estimates have been made of the gas leakage rate in natural gas production and distribution, but they vary widely between countries and sources. A rate of 0.5% has been assumed for this analysis. Although this is below many estimates, it is intended to represent the marginal leakage due to additional gas demand. New pipes should have a low leakage rate. The marginal effect of new gas demand on leaks from old residential distribution pipes may be negative.

Tailpipe emissions of methane could be much higher than those assumed here. Existing CNG cars have methane emissions in the region of 1g/km, adding around 10% to the life-cycle greenhouse gas emissions.

Greenhouse gas emissions due to power generation for gas compression are significant, contributing over 10% of the life-cycle emissions in the North American case, though rather less in Europe.

6.3.5   Alcohol Fuels

This report considers two alcohol fuels: methanol and ethanol. Both can be derived from a variety of sources. For methanol the sources of interest are natural gas, coal and biomass. All can be converted to synthesis gas, and hence to methanol. As the main concern of this study is greenhouse gas abatement, production of methanol from coal is not covered in any detail — use of this process would result in higher CO_2 emissions than arise from gasoline use.

Production. Ethanol can be produced from any biomass feedstock containing simple sugars, starch or cellulose. This study considers the production of ethanol from maize, wheat and wood.

While biomass feedstocks can be produced on a renewable basis, production volumes are limited by competing land-use applications. Recent studies indicate that no more than 15% of transport fuel demand in the United States could be met from wood-based methanol production (OECD, 1993). Estimates of biomass availability in the European Community give comparable results (CEC, 1992).

Ethanol can be produced from wheat, maize and other starchy crops by fermentation using yeast. The crop is cooked and mashed in water, and the starch is hydrolysed to sugar using the enzyme amylase. The sugar is then fermented to give alcohol. The alcohol is separated from the water, yeast and other residues by distillation. The residues, including spent yeast, are extracted and used as animal feed. Several variations on the process are possible, involving batch or continuous fermentation and separation of the animal feed by-product at different stages. The economics of all the processes depend heavily on the value of the by-product, especially where wheat is the feedstock.

Overall energy use in the process ranges from 30% to 60% of the energy content of the ethanol produced (Marrow *et al.*, 1987). Ethanol costs depend on the feedstock price and on revenue from by-product sales. A metric ton of wheat yields about 0.29 metric ton of ethanol, so with a wheat price of $120 a ton the feedstock component of the cost is over $0.30 a litre.

Crop residues such as corn stover and wheat straw can in principle be used as process fuel although it may be preferable to dig them in as fertiliser. Some nutrients are retained in ash when the residues are burned. Others, such as nitrogen, are lost.

Wood can also be used to produce ethanol. The main constituents of wood are lignin, cellulose and hemicellulose. Lignin is a tarry substance which must be separated from the other components to allow them to be hydrolysed and fermented to ethanol. Several processes involving acid treatment or enzymes are under development, but none is near commercialisation. The energy required to make ethanol from wood could be one and a half to three times the energy content of the ethanol produced.

Methanol production from wood probably has greater long-term potential than that of ethanol from food crops. The wood is gasified and the gases cleaned, producing synthesis gas (a mixture of carbon monoxide and hydrogen), which is converted to methanol. The overall process efficiency is expected to be about 55%.

The conversion of wood to methanol can lead to greenhouse gas emissions from electricity use, mainly to produce oxygen where the gasifiers are oxygen-blown. Greenhouse gases can also be emitted in forestry.

Storage and Handling. Methanol has half the volumetric energy density of gasoline, and ethanol about 60% (USDOE, 1990a). The lower energy density affects the entire fuel cycle, including fuel storage and distribution from the point of production. The whole vehicle system has to be modified, including a larger fuel tank for the same range as a gasoline vehicle. For gasoline-alcohol flexible fuel vehicles, certain components must allow highly variable fuel flow rates.

Alcohol fuels, methanol in particular, corrode some materials currently used in engines, including rubber and steel. Materials in fuel tanks, fuel lines and fuel injection systems must be made of appropriate materials.

When ingested or absorbed through the skin, methanol is more toxic than gasoline. The safety problem is aggravated by the fact that methane vapour is virtually odourless. As alcohol is water soluble, leaks from storage tanks could seriously contaminate groundwater. Methanol or ethanol used as a transport fuel must be denatured to give an odour and flame colour. Vapour accumulation in storage facilities could also present problems.

Engine. Alcohol fuels have high octane ratings (105 for M85, 110 for M100[1]), making it possible to increase compression ratios to about 12:1 compared with the norm of 9:1 for gasoline engines. The high latent heat of methanol vaporisation also provides cooling of the engine charge, reducing compression work, improving engine efficiency and potentially reducing NO_x emissions. Broader flammability limits allow for later ignition, yielding a 5-7% energy efficiency gain over regular grade gasoline and 2-4% over premium. An overall efficiency gain between 10% and 20% is possible in principle with M100, compared with an equivalent gasoline engine, though this has not been demonstrated. The presence of an oxygen atom in the alcohol molecule aids complete combustion, reducing carbon monoxide and VOC emissions.

---

1. M85 is 85% methanol and 15% gasoline; M100 is pure methanol.

Cold-start problems can be caused by the high latent heat of vaporisation of alcohol fuels. These are aggravated in cold weather. Solutions add to the total cost of the vehicle (USDOE, 1990b). Volatile starting aids (gasoline, propane, ether, etc.) or an auxiliary device are required at temperatures below about 10°C with classical spark-ignition engines. In very hot climates, vapour lock can be a problem as the boiling point of methanol is well below that of gasoline, although Ford has solved this with an extra fuel pump (Heath, 1991).

Diesel engines can be adapted to run on ethanol or methanol. Fuel economy is about 5% better than with diesel fuel. It is necessary to add ignition and lubricity enhancers to the fuel, increasing its cost by about 20%. Higher capacity fuel pumps and injectors must also be used, adding around $500 to the engine cost. No other engine changes are needed (CEC, 1992).

Life-cycle Emissions. Only unblended alcohol fuels (ethanol and methanol) are considered in this analysis. Mixtures of methanol and gasoline, such as M85, are excluded. The alcohol is assumed to be burned in an Otto-cycle engine. As methanol and ethanol have roughly the same properties, the engine characteristics are assumed to be the same. Ethanol has a slightly higher energy density, however, so the fuel system is lighter than for methanol.

Methanol from natural gas has life-cycle greenhouse emissions similar to those of gasoline (4% lower in North America and 7% lower in OECD Europe). As **Figure 6.5** shows, emissions from the tailpipe are 20-23% below the level for gasoline vehicles. Emissions from fuel production and distribution are roughly 60% higher.

Figure 6.5
Methanol Car Life-cycle Greenhouse Gas Emissions
(Methanol from Natural Gas)

Based on North American vehicles and energy supply system. Life-cycle emissions for new cars in OECD Europe in 2000 are not expected to be significantly different.

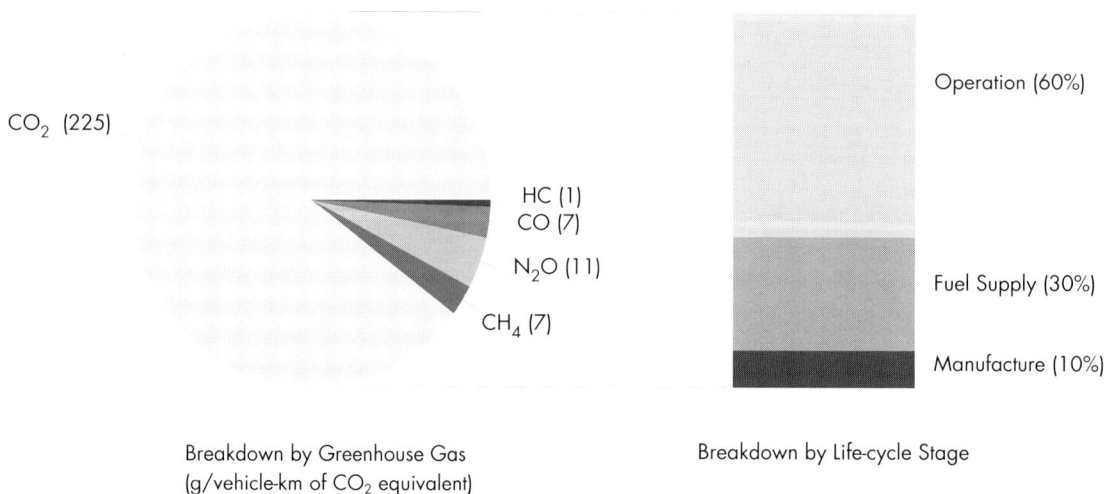

CO_2 (225)

HC (1)
CO (7)
N_2O (11)
CH_4 (7)

Operation (60%)
Fuel Supply (30%)
Manufacture (10%)

Breakdown by Greenhouse Gas
(g/vehicle-km of CO_2 equivalent)

Breakdown by Life-cycle Stage

Three bio-alcohols have been considered in the emissions model: ethanol produced from maize, ethanol produced from wood and methanol produced from wood. The CO_2 emissions from vehicles using them are treated in the model as being zero or negative because of CO_2 taken up by the biomass during growth. The remaining greenhouse gas emissions from the vehicle are mainly N_2O, carbon monoxide and VOCs. Other CO_2 and non-CO_2 greenhouse gas emissions may arise from fertiliser production and the use of fossil fuels in farm equipment. The emissions model shows methanol and ethanol from lignocellulosic material to have life-cycle greenhouse emissions 70% lower than those for gasoline.

Maize-derived Ethanol. About 84% of life-cycle greenhouse gas emissions in the maize-ethanol fuel cycle arise during fuel supply (see **Figure 6.6**). CO_2 emissions are produced from energy use during fertiliser manufacture and farm equipment operation. N_2O is also emitted in significant quantities through denitrification of nitrates in the soil. The quantity of N_2O produced is highly uncertain; the emission factors used are based on a number of measurements and estimates (DeLuchi, 1992, Volume 2, Appendix N). N_2O emissions associated with wheat production are thought to be smaller, perhaps by a factor of three.

In this analysis, an energy mix of 40% solid fuel, 40% natural gas, 10% electricity and 10% biomass is used in the fermentation and distillation processes. This accounts for 43% of life-cycle greenhouse gas emissions in North America and roughly 37% in OECD Europe. Using only coal to fire the distillation boilers (as is generally the case in the United States) somewhat increases overall emission levels. The use of biomass would dramatically reduce them.

Figure 6.6
Ethanol Car Life-cycle Greenhouse Gas Emissions
(Ethanol from Maize)

Based on North American vehicles and energy supply system. Life-cycle emissions for new cars in OECD Europe in 2000 are not expected to be significantly different.

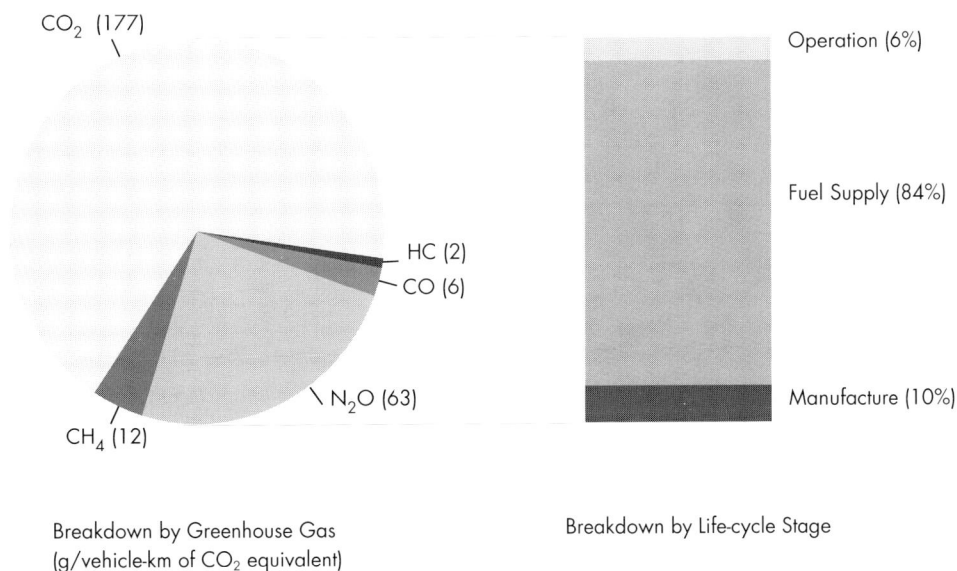

CO_2 (177)

HC (2)
CO (6)

N_2O (63)

CH_4 (12)

Operation (6%)

Fuel Supply (84%)

Manufacture (10%)

Breakdown by Greenhouse Gas
(g/vehicle-km of CO_2 equivalent)

Breakdown by Life-cycle Stage

Wood-derived Methanol. Figure 6.7 shows the life-cycle emissions from the use of methanol from wood. Vehicle manufacture is very important, contributing 30% of life-cycle emissions, as net emissions from fuel use are very low. The greenhouse gas emissions arising from this fuel cycle depend on the degree to which fertilisers and lime are used in silviculture, as well as the need for irrigation. This study assumes that no artificial irrigation or lime is used, and that only low levels of fertiliser are needed, on no more than about half of the cultivated terrain. Despite low fertiliser use, N_2O is the main non-CO_2 greenhouse gas. Wood is assumed to be dried on site with waste heat before processing. Energy requirements for planting, harvesting and chipping are met by diesel fuel.

Figure 6.7
Methanol Car Life-cycle Greenhouse Gas Emissions
(Methanol from Wood)

Based on North American vehicles and energy supply system. Life-cycle emissions for new cars in OECD Europe in 2000 are not expected to be significantly different.

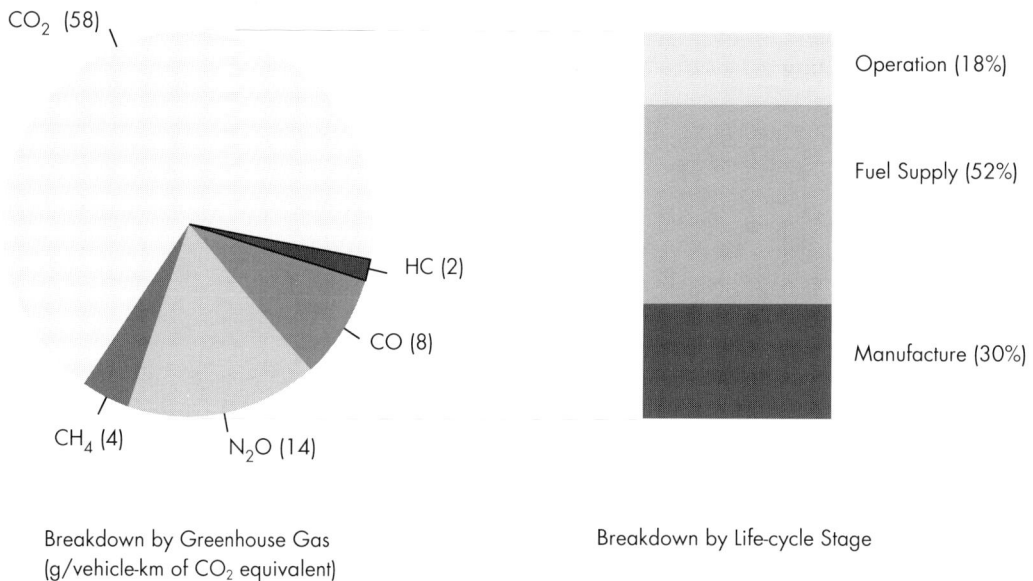

CO$_2$ (58)
HC (2)
CO (8)
CH$_4$ (4)
N$_2$O (14)

Operation (18%)
Fuel Supply (52%)
Manufacture (30%)

Breakdown by Greenhouse Gas
(g/vehicle-km of CO$_2$ equivalent)

Breakdown by Life-cycle Stage

The conversion of wood to methanol leads to greenhouse gas emissions from the consumption of electricity and from the gasification and combustion of wood. The most important of these is likely to be electricity use for oxygen plant.

6.3.6 Hydrogen

In this analysis hydrogen is assessed as a fuel for an Otto-cycle engine. Use of hydrogen in fuel cells is not considered. Hydrogen is included in the analysis to illustrate the long-term potential for emission reductions. It is not expected to be used widely in vehicles before 2010. The process examined here is far from economic.

Production. Any energy vector, including fossil fuels, biomass and electricity, can be used to produce hydrogen. Hydrogen may be produced from water by thermal or thermochemical conversion, photolysis or electrolysis. Hydrogen can also be produced on board vehicles by reforming methanol or other fuels.

Most hydrogen is currently produced from natural gas by reformation at efficiency of around 70-75% (IEA Coal Research, 1984). This is the cheapest, most efficient means of hydrogen production from a primary energy source. Hydrogen could be produced by gasifying coal or biomass, with efficiency of 60-65%. The cost would be lower than that of methanol from the same source.

Electrolysis of water can be a clean means of hydrogen production, but it is expensive. If the local electricity supply is fossil-based, it would be more efficient to produce hydrogen directly from the fuel by thermochemical methods.

The hydrogen fuel cycle considered here is arranged to minimise emissions. Hydrogen produced by electrolysis using non-fossil electricity is the method considered because it is potentially the cleanest production technology. Coal has been assumed to be used in parts of the electricity supply industry, resulting in some greenhouse gas emissions.

Electrolysis of water has an energy efficiency level of 75% using what is currently the best commercial technology. An efficiency level of 83% is assumed for future technology. Hydrogen is assumed to be transported in gaseous form by pipeline from point of production to point of use, and liquefied before use. The pipeline is assumed to be leakproof. The energy consumed by pipeline compressors is assumed to be hydrogen. Non-fossil power is assumed to be used to drive small-scale hydrogen liquefiers at fuelling stations.

Storage. Hydrogen has very high energy content per kilogram, but very low volumetric energy density. Therefore the fuel tank tends to be heavy and bulky, and the use of hydrogen for transport presents a major technological challenge.

Hydrogen fuel can be stored in the vehicle in gaseous form in pressure vessels at about 200 bar, as a cryogenic liquid at -253°C in a dewar vessel, or chemically bound in metal hydride storage. Hydrogen has as yet been used as a vehicle fuel only in experimental situations. All three means of storage have been used and no clear advantage to any one has been demonstrated.

High-pressure storage is simple but heavy and bulky. The energy equivalent of 75 litres of gasoline is supplied by about one cubic metre of hydrogen at 225 bar. The cylinder weight would be about 1 500 kg. This weight and volume are not acceptable in a passenger car and vehicle range would be reduced.

There are great difficulties in distribution and storage of liquid hydrogen. Some of the hydrogen evaporates even from the best insulated storage containers. The losses can amount to 0.5-3% of the fuel stored in the tank per day, creating a potential safety hazard. More fuel is lost during transfer (between 10% and 25% is boiled off during refuelling).

The weight of a hydride storage system can be roughly the same as for compressed gas cylinders but it takes up almost one-third more space. Hydride tanks are easier to fit on a car than high-pressure cylinders because they can be shaped to fit available space. Although storage has been demonstrated using a titanium-iron hydride, the system is far from ready for commercialisation.

Engine. The most attractive engine for hydrogen may be a spark-ignited, direct-injection engine. Such engines are relatively fuel-efficient irrespective of the fuel used. In the long term, however, fuel cells may be the most attractive means of propulsion using hydrogen.

The very low ignition energy and wide flammability limits of hydrogen cause some problems in an Otto-cycle engine. Special measures are needed to regulate combustion in order to avoid backfiring, pre-ignition and knock. Backfiring — combustion in the intake manifold — causes a loss of power and damages the engine. Solutions that have been used include water injection to suppress ignition, exhaust gas recirculation and very lean-burn.

The wide flammability range of hydrogen allows operation with a very lean air/fuel mixture. Daimler-Benz has experimented with hydrogen in converted gasoline engines, at an equivalence ratio of $\lambda \approx 3$. In such lean conditions knock and backfiring do not occur, and the very low flame temperature results in zero NO_x emissions.

Life-cycle Emissions. Figure 6.8 shows life-cycle emissions from a hydrogen-powered car. While no CO_2 emissions arise at the tailpipe from the combustion of hydrogen, the combustion of lubricants produces small amounts of CO_2 and carbon monoxide. The remaining greenhouse emissions for the hydrogen fuel cycle arise in production and distribution and in manufacturing the car. Most of the emissions stem from energy use in the electricity industry.

Figure 6.8
Hydrogen Car Life-cycle Greenhouse Gas Emissions
(Hydrogen from Water, Electrolysed with Nuclear Power)

Based on North American vehicles and energy supply system. Life-cycle emissions for
new cars in OECD Europe in 2000 are not expected to be significantly different.

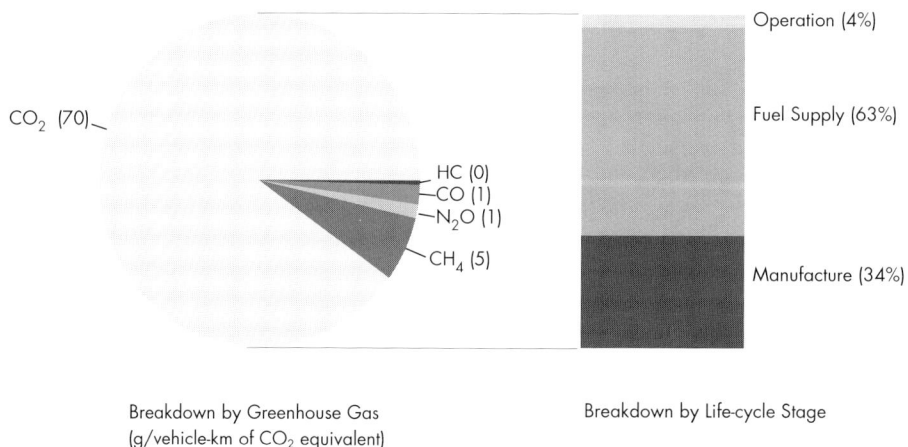

CO_2 (70)

HC (0)
CO (1)
N_2O (1)
CH_4 (5)

Operation (4%)

Fuel Supply (63%)

Manufacture (34%)

Breakdown by Greenhouse Gas
(g/vehicle-km of CO_2 equivalent)

Breakdown by Life-cycle Stage

6.3.7 Electric Vehicles[1]

Electric cars were first produced in the late 19th century but production ceased around the 1930s as internal combustion engine technology rapidly improved. Their disadvantages have generally included high cost, limited range and low power (limiting acceleration and maximum speeds), as well as heavy, voluminous batteries. Nevertheless they are used for some applications where noise and gaseous emissions are unacceptable and only low speeds are required.

Improvements in battery and electric motor technology are beginning to narrow the very large performance gap between electric vehicles and gasoline or diesel light-duty vehicles. Interest has been stimulated in recent years by concern about urban air pollution.

Government policy favouring electric vehicles has elicited a variety of development programmes, including the US Advanced Battery Consortium, the plan by Japan's Ministry of International Trade and Industry to promote development and use of electric vehicles and Calstart, the California Electric Car Consortium. Most of the development work focuses on improving drive-train efficiency and raising batteries' energy storage density.

Electricity Storage. If the electric vehicle is to be competitive, better batteries or another, more efficient power source must be developed. R&D is being conducted on ten to twenty different types of batteries. The aims of the R&D include improving specific power and energy storage and reducing costs, maintenance needs and recharging times.

The electric vehicles currently on the market use lead/acid batteries. The lead/acid battery has limited specific energy (around 20-25 Wh/kg in normal driving) and is consequently very large and heavy, but has a satisfactory power density. Maintenance requirements are generally acceptable, although some versions of these batteries require periodic watering, which increases vehicle maintenance costs. Lead/acid batteries are less expensive than the current alternatives.

Maintenance-free lead/acid batteries are being developed and test results indicate that their performance is better than that of conventional lead/acid batteries (DeLuchi et al., 1989). However, these batteries are unlikely to match the levels of specific energy attained by other batteries under development. The near-term alternatives are nickel-cadmium (Ni/Cd), sodium-sulphur (Na/S) and nickel-iron (Ni/Fe) batteries. The Ni/Cd battery is already commercialised and the others are expected to follow within a few years. These batteries, all providing higher specific energy than the conventional lead/acid battery, are likely to be used in prototype cars during the 1990s.

Electric Motors. Most existing electric vehicle prototypes have direct current (DC) commutated motors. This motor design is mature, but small improvements can still be made with regard to weight, compactness, efficiency and maintenance requirements. More significant improvements can be made to motor controllers. Both motor and controller efficiency is lowest at high torque and low speed. In urban driving, motor/controller efficiency can be as low as 50%.

---

1. Further information on this subject is available in two IEA/OECD publications — the forthcoming *Current Status of Electric Vehicle Research and Development* (Draft Final Report, May 1992) and *The Urban Electric Vehicle: Policy Options, Technology Trends and Market Prospects* (OECD, Paris, 1992).

Alternating current (AC) induction motors do not have commutators or rotor windings. They have a simpler structure and require less maintenance than DC commutated motors. They are also 50% lighter and 75% cheaper. In the past they were not considered for electric vehicles because of the difficulty of converting a DC battery output to a variable frequency AC motor input. However, recent advances in power electronics are such that AC motor controllers are not much more expensive or heavy than DC controllers. AC motors are therefore expected to be used in most new electric vehicle prototypes in the mid- to late 1990s.

The torque obtained from AC motors is less speed-dependent than that from DC motors. Whereas it has been necessary to use gearboxes with at least two speeds for cars powered by DC motors, AC motors are expected to be able to operate with single-speed transmissions.

Life-cycle Emissions. A separate comparison has been made between electric vehicles and other vehicles under urban driving conditions. This reflects the limited market in which electric vehicles would be able to compete with internal combustion engine vehicles.

Electric vehicle performance is based on the use of a sodium-sulphur battery with energy density of 120 Wh/kg, 75% discharge efficiency and 92% charging efficiency. The electric vehicle is assumed in the model to have an urban range of 240 km and to weigh 295 kg (roughly 25%) more than the gasoline vehicle. The weight increase takes account of structural reinforcement required to carry the batteries.

The electric vehicle is assumed to have roughly six times the efficiency of the equivalent internal combustion vehicle (in km/GJ at the battery terminals versus km/GJ from the fuel tank) for a given level of performance, net of the weight difference. This corresponds to an electric motor/controller and drive-train efficiency of about 70%.

It is expected that owners of electric vehicles would charge their batteries at night, when inexpensive baseload power is available. Accordingly, the use of electric vehicles would help level the load profile for power generators[1]. This would have a different impact according to the local generation mix. The power plants used to charge electric vehicles would in general be fired by coal, oil or gas, since nuclear and hydro plants are already run at maximum load in most countries. Although night charging of electric vehicles would be the ideal situation, it is likely that some daytime charging will occur.

Calculations have been made of emissions from electric vehicles, with batteries charged from several power generation mixes. **Figures 6.9a** and **6.9b** show life-cycle emissions arising from electric vehicles recharged from the average regional power mix, coal-fired generation, oil-fired, combined cycle gas-fired, nuclear power, and hydropower or other renewables. These are compared with unleaded gasoline, diesel and CNG vehicles.

The two extreme emission cases are coal-fired generation and hydropower. Electric vehicles using electricity from coal have higher life-cycle emissions than gasoline vehicles. Electric vehicles using hydroelectricity have lower emissions than any other technology. Electric vehicles in OECD Europe have lower emissions than those in North America because of the higher share of nuclear power. The power generation characteristics in each region are described in **Annex 3**.

---

1. It is assumed that the demand for electric vehicles would not be so great as to completely offset the day-to-night load drop. The small expected load increase reflects the small market expected for electric vehicles.

94

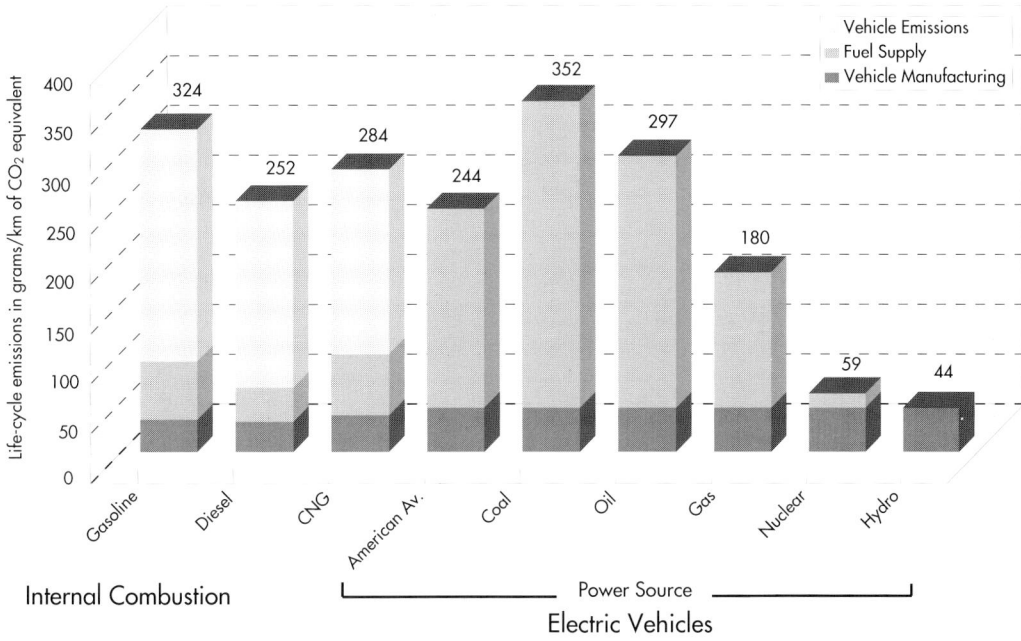

Figure 6.9a
Electric Vehicle Life-cycle Greenhouse Gas Emissions, North America, 2000
(grams/km of CO_2 equivalent in urban driving)

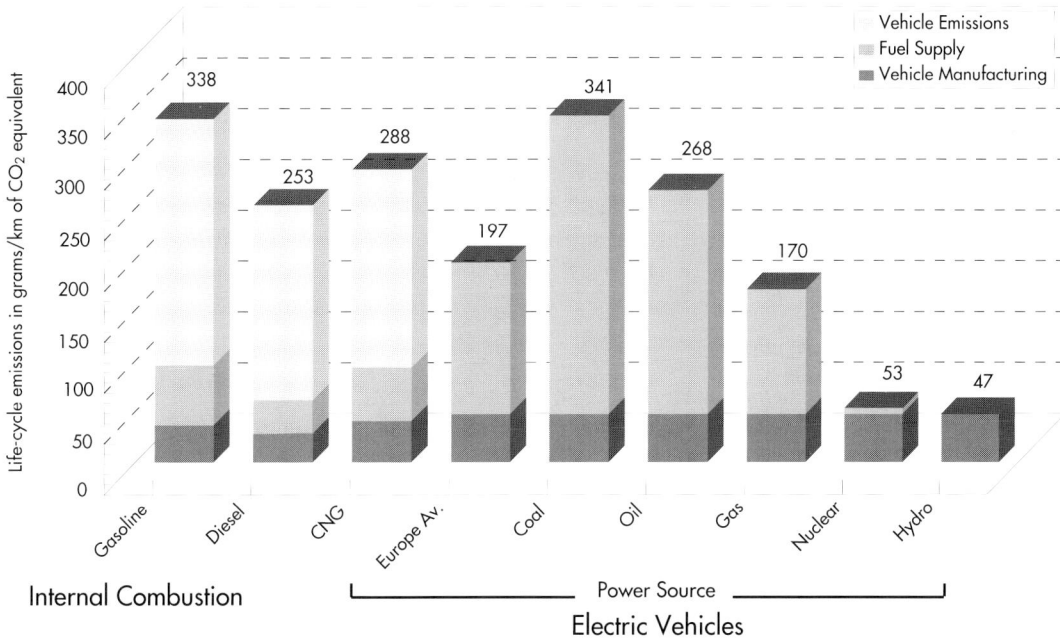

Figure 6.9b
Electric Vehicle Life-cycle Greenhouse Gas Emissions, OECD Europe, 2000
(grams/km of CO_2 equivalent in urban driving)

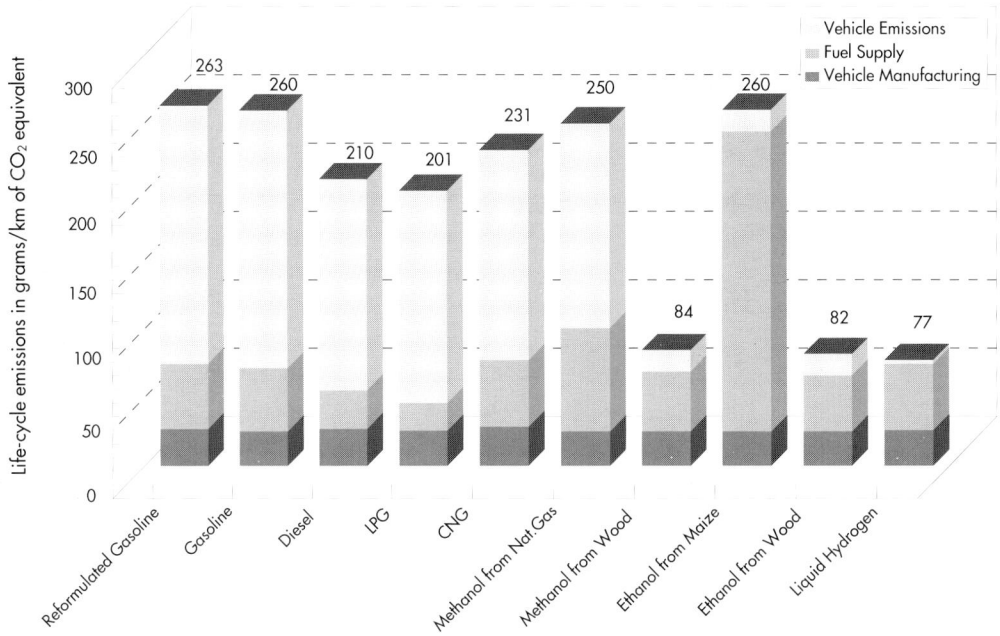

Figure 6.10a
Life-cycle Greenhouse Gas Emissions for Alternative-Fuel Cars, North America, 2000
(grams/km of CO_2 equivalent)

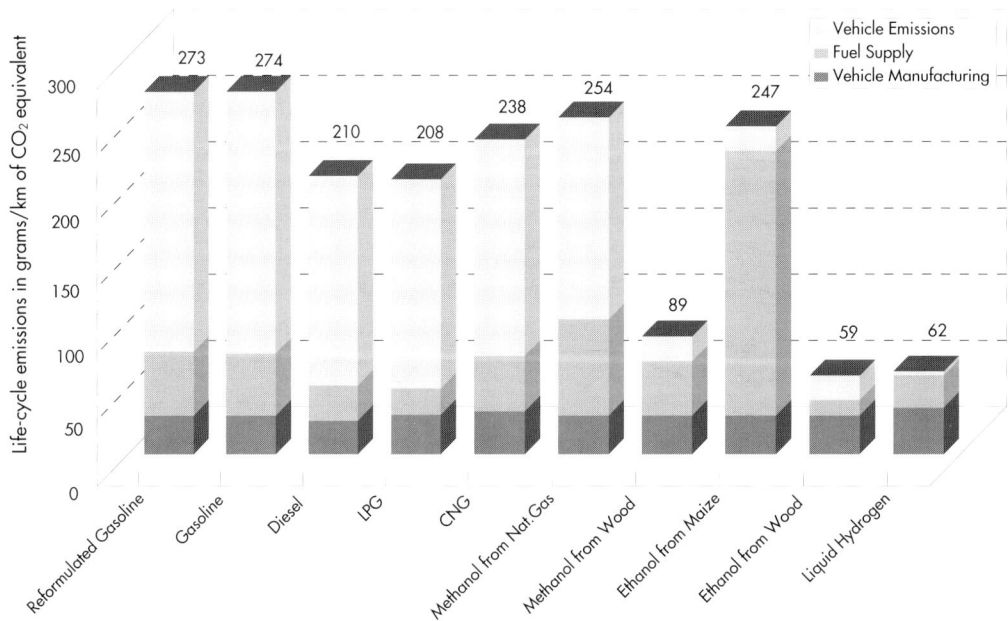

Figure 6.10b
Life-cycle Greenhouse Gas Emission for Alternative-Fuel Cars, OECD Europe, 2000
(grams/km of CO_2 equivalent)

6.4 Comparison of Greenhouse Gas Emissions

Life-cycle emissions from some of the options considered in this chapter are compared in **Figure 6.10a** and **6.10b**. Four groups of vehicle/fuel combinations emerge from the analysis:

- conventional light-duty vehicles using gasoline;
- market-ready options with moderately reduced emissions: CNG, LPG and diesel;
- pre-commercial options with low greenhouse gas emissions: liquid hydrogen produced by electrolysis of water using non-fossil electricity, electric vehicle with non-fossil electricity, biofuel with low fossil inputs;
- pre-commercial options with high greenhouse gas emissions: electric vehicles or hydrogen based on fossil-fuel electricity, methanol from coal or natural gas, ethanol from cereals using high fossil inputs.

Although it would be possible to rank the vehicle options based on the analysis in this chapter, the ranking would probably not be valid for any one country. As **Figure 6.11** shows, the life-cycle emissions of the various alternative vehicles depend strongly on the assumptions used in the model.

Figure 6.11
Greenhouse Gas Emission Ranges for Alternative-Fuel Cars
(Reformulated Gasoline = 100)

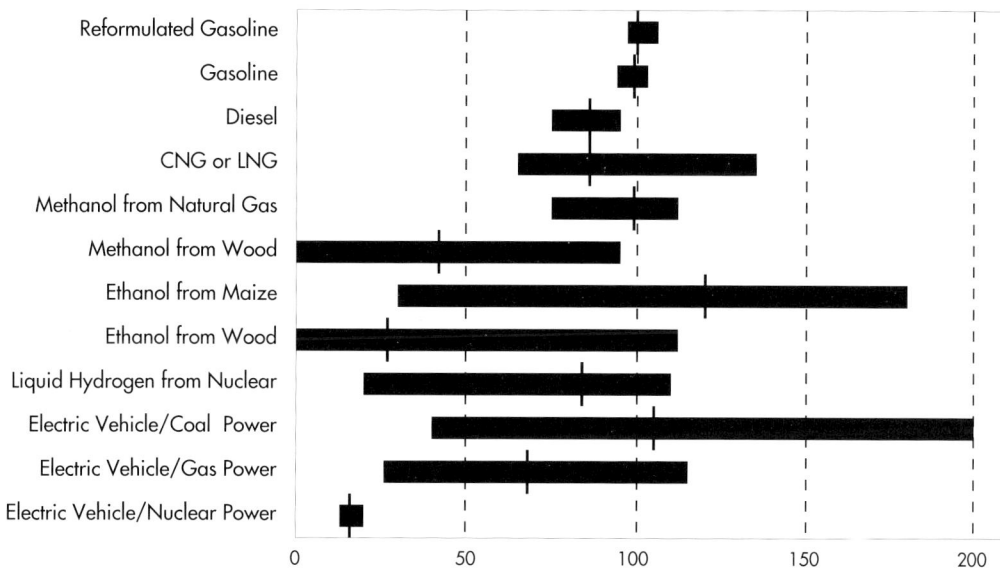

Note: This graph indicates the possible range of greenhouse gas emissions per kilometre for each alternative fuel and for electric vehicles with various power sources. The ranges can result both from uncertainties about the car technologies and from local variations between energy supply technologies and systems.

Life-cycle Analysis: Future Needs. The analysis presented in this chapter produces results comparable with other life-cycle analyses of transport fuels (for example, CEC, 1992; DeLuchi, 1992; Ecotraffic, 1992). Further work is needed to standardise the approaches used for life-cycle analysis and to collate the life-cycle emission inventories that are being generated in many countries.

CHAPTER 7

ECONOMIC POTENTIAL OF ALTERNATIVE FUELS

Life-cycle analysis of the greenhouse gas emissions from the use of cars and fuels is the first step in appraising emission management options. This chapter aims to provide a further step, by considering the cost-effectiveness of alternative fuels for greenhouse gas abatement. This step is by no means the last in developing policy strategies. Governments have a large number of political aims to balance and may wish to promote alternative fuels that would not have been chosen purely for cost-effective greenhouse gas abatement. Conversely, fuels that appear cost-effective for greenhouse gas abatement may be unattractive for other reasons, including their environmental effects. These issues are considered further in **Chapter 11**.

Many of the fuels considered in **Chapter 6** are likely to be more expensive than gasoline. The IEA has estimated cost ranges for most of them (IEA, 1990b). **Table 7.1** shows the life-cycle greenhouse gas emissions per kilometre for several fuels, calculated for North America in **Chapter 6**. It also shows estimates, based on the assumed fuel economy levels, of the fuel costs to drivers excluding tax. From these data a first estimate of the cost of greenhouse gas reduction can be obtained. **Figure 7.1** shows the result of this calculation. The graph shows the extra fuel cost per metric ton of CO_2-equivalent not emitted by switching from gasoline to each of several fuels. Methanol from natural gas and ethanol from maize and wheat are not shown, as they have very limited potential to reduce greenhouse gas emissions.

Table 7.1
Substitute Fuels for Cars: Costs and Greenhouse Gas Emissions

Fuel	Life-cycle Emissions CO_2 Equivalent (g/km)	Energy (MJ/km)	Fuel Cost (US 1987 ¢/km)
Gasoline	260	2.3	1.7 - 2.1
Diesel	210	2.0	1.3 - 1.7
CNG	231	2.2	0.9 - 2.0
Methanol from Natural Gas	250	2.0	1.1 - 2.4
Methanol from Wood	84	2.0	2.3 - 4.6
Ethanol from Maize and Wheat	260	2.0	2.4 - 3.7
Ethanol from Wood	82	2.0	1.8 - 8.5
Liquid Hydrogen from Electicity	77	1.9	2.8 - 5.6

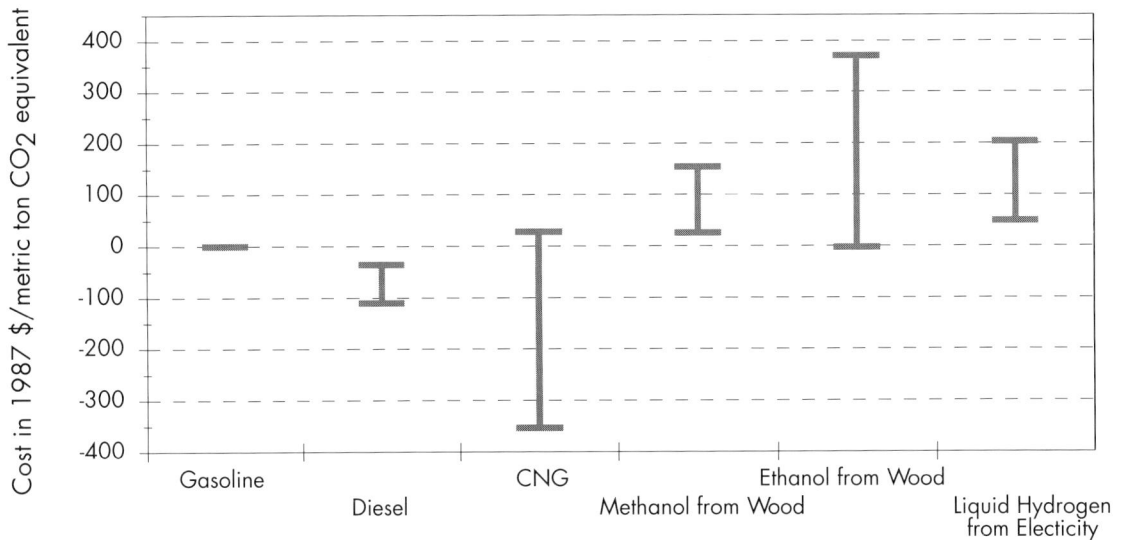

Figure 7.1
Alternative Fuels' Cost-Effectiveness for Greenhouse Gas Abatement
(1987 US $ per metric ton of CO_2 equivalent not emitted)

Sources: IEA (1990b), Marrow and Coombs (1990), CEC (1992).

The graph indicates that both diesel and CNG may result in lower emissions at lower cost. This means they may have a negative cost of abatement, though costs of car operation are not included here. Despite their limited emission reduction potential, this chapter concentrates on diesel and CNG, as they appear to be the most financially attractive way of beginning to reduce emissions and they are already commercially available in the OECD. A detailed examination is made of the economics of gasoline, diesel and CNG cars using a spreadsheet-based discounted cash-flow analysis.

7.1 UNDERLYING ASSUMPTIONS

The analysis was carried out using car and fuel costs typical of two OECD countries: the United States and France. The choice of countries provides an interesting contrast in the types of car being used, the prices paid for them and the costs of operating them. There are also significant differences in government policies affecting the car markets in the two countries.

In France, cars are lighter than those in the United States and their fuel consumption is about 35% lower. Car prices are about the same in both countries, but fuel prices in France are much higher. France has lower tax on diesel than on gasoline, while the United States has recently introduced legislation providing tax credits for the use of alternative fuels, including CNG.

The analysis is set in 2000, by which time production-line CNG cars may be available and the direct-injection diesel car is likely to be more widespread. Emission standards in France and the United States have differing effects on the type of diesel engine technology that can be used.

Car Costs. In an attempt to make the analysis representative for France and the United States, car purchase and operating costs are estimated starting from typical costs for 1992[1]. **Tables 7.2** and **7.3** show the assumed purchase and operating costs and energy economy of cars in each country in 2000. Gasoline cars are assumed to have the 1992 fuel economy level for the models considered and sensitivity checks have been carried out for a 10% improvement. This range is consistent with a continuation of the fuel economy trends during the late 1980s in France and the United States.

Table 7.2
Cars in the United States in 2000
Assumptions for Base Case

	Gasoline	Diesel	CNG (2/3 range)	CNG (full range)
Pre-tax Vehicle Price ($)	17 000	17 850-19 550	17 677-18 403	18 080-19 210
Post-tax Car Price ($)	18 020	18 921-20 723	18 737-19 507	19 165-20 362
Durability (km)	200 000	250 000	200 000	200 000
Maximum Age (years)	10	10	10	10
Pre-tax Price of Fuel, $/litre gasoline equivalent	0.24	0.20	0.13-0.20	0.13-0.20
Fuel Tax $/litre gasoline equivalent	0.10	0.12	0.10	0.10
Fuel Retail Price, $/litre gasoline equivalent	0.35	0.32	0.24-0.30	0.24-0.30
Fuel Consumption (litres gasoline equiv./100 km)	11.76	9.41	11.29	11.70
Other Running Costs: Distance Between Services, @$160: (km)	10 000	6 667	15 000	15 000
Distance Between Tyre Changes, @$235: (km)	40 000	38 000	36 534	34 478
Repairs, $ per 10 000 km	360	378-414	375-390	383-407
Cost of Catalyst (replaced every 100 000 km) ($)	800	800	800	800
Annual Licence Fee ($)	36	36	36	36
Annual Insurance Premium ($)	885	903-939	900-915	908-932

Note: Vehicle based on 1992 Ford Taurus with 3 litre engine. Prices are based on second quarter 1992.
         Vehicle costs and fuel consumption based on information supplied by Runzheimer (1993).

1.   The analysis for the United States is based on information from Runzheimer International (1993), an organisation that carries out surveys of car user expenditure. The data are comparable with those from the American Automobile Association presented in Davis & Morris (1992). The analysis for France is based on survey results from the National Automobile Club and from Runzheimer International.

Table 7.3
Cars in France in 2000
Assumptions for Base Case

	Gasoline	Diesel	CNG (2/3 range)	CNG (full range)
Pre-tax Vehicle Price (FF)	79 000	79 000-90 850	81 253-83 767	82 638-86 536
Registration (FF)	516	516	516	516
Retail Vehicle Price (FF)	97 010	97 010-111 467	99 758-102 825	101 448-106 204
Durability (km)	160 000	200 000	160 000	160 000
Maximum Age (years)	10	10	10	10
Pretax Price of Fuel, FF/litre gasoline equivalent	1.36	1.24	0.92-1.19	0.92-1.19
Fuel Tax, FF/litre gasoline equivalent	3.65	1.74	3.65	3.65
Fuel Retail Price FF/litre gasoline equivalent	5.01	2.98	4.42-4.75	4.42-4.75
Fuel Consumption (litres gasoline equiv/100 km)	7.60	6.08	7.27	7.52
Other Running Costs: Distance Between Services @ 160FF: km	10 000	6 667	15 000	15 000
Distance Between Tyre Changes @1500FF: km	40 000	38 000	36 760	34 831
Repairs, FF per 10 000 km	1 940	1 940-2 229	1 995-2 056	2 029-2124
Cost of Catalyst (replaced every 100 000 km) (FF)	4 000	0-4 000	4 000	4 000
Annual Licence Fee (FF)	447	447	447	447
Annual Insurance Premium (FF)	3 425	3 425-3 787	3 494-3 571	3 536-3655

Note: Vehicle based on Renault Clio with 1.4 litre engine. Prices are based on second quarter 1992. $1 = 5.2 FF
Vehicle costs and fuel consumption based on information supplied by Runzheimer (1993) and the French National
Automobile Club (Automobile Club National).

The diesel car for both countries is assumed to have direct fuel injection, and to use 20% less fuel than the gasoline car in energy terms (about 30% by volume). This is the basis on which greenhouse gas emissions are calculated in **Chapter 6**. The energy use by the CNG car is calculated assuming 10% higher engine efficiency, which again was the assumption made in **Chapter 6**. Allowance is made for a fuel consumption penalty for CNG, due to the weight of storage cylinders.

The costs of diesel and CNG cars relative to gasoline cars are not predictable, even in the very short term, so a range of costs has been considered in each case for both countries. Prices depend on the interaction between consumer demand for the cars and the timing and scale of manufacturers' investment in new plant to produce them. Diesel cars were cheaper than gasoline cars of the same model in some markets in 1992, even though manufacturing costs were almost certainly higher.

Performance. Gasoline and diesel engines differ in the profile of maximum torque with engine speed. This makes it hard to define "equivalent" performance. A diesel car can have higher acceleration from standstill, but lower maximum speed, than a gasoline car. The issue is partly avoided here by the use of ranges of vehicle costs.

CNG cars are unlikely to offer the same benefits to their owners as gasoline cars. CNG storage is bulky, and purchasers may have to decide between a shorter driving range or reduced cargo space. Where a CNG car is required to have the same driving range as its gasoline equivalent, its cost is likely to be much higher.

Gasoline Cars. The average purchase price of a gasoline car increased in the United States at a rate of 1.5% a year in real terms between 1970 and 1990 (Davis and Morris, 1992). This trend is assumed to continue during the 1990s. In France, gasoline car prices are assumed to rise in real terms at a higher rate of 2.3% because catalytic converters were being introduced in Europe during the early 1990s. This growth rate is similar to that in the United States during the 1980s following the introduction of emission standards requiring catalytic converters.

Diesel Cars. The cost of diesel cars in the coming decade will depend partly on the emission standards they are required to meet. EC standards in force from 1992/93 can be met by existing diesel cars. The standards proposed for 1996 and 1999 require lower emissions of NO_x and particulates, but they are probably achievable without a significant change in diesel engine technology. US federal standards for the mid-1990s will require more advanced technology.

The analysis was carried out for diesel car prices 5% to 15% higher than those of gasoline cars in the United States: 5% is probably representative of the real cost of replacing gasoline engine components with diesel engine components. However, shorter production runs for diesel engines, as well as the need to amortise development costs, are likely to raise the prices sought by manufacturers.

If diesel cars can meet EC standards in 2000 with minimal modifications, they may be no more expensive than gasoline cars fitted with three-way catalytic converters. For France the analysis has therefore been carried out for a range of diesel car prices from zero to 15% above those of gasoline cars.

CNG Cars. The excess cost of a CNG car over a gasoline car is assumed to be determined by the storage equipment cost. The difference in the costs of components for gaseous fuel engines is small compared with the uncertainty in cylinder costs. Mass-produced CNG cars have been predicted by the EPA to cost $900 more than equivalent gasoline cars, while work by the World Bank indicates cost differences around $1 000 (Sypher:Mueller, 1992). CNG cars with two-thirds of the gasoline car driving range have been considered, as well as with the full gasoline range. Costs, weights and fuel consumption levels have been calculated as shown in **Tables 7.4** and **7.5**.

Table 7.4
CNG Car Additional Cost and Weight
United States

Gasoline Car Assumptions		
Range on Full Tank, km	510	
Loaded Vehicle Weight, kg	1 750	
Gasoline Weight (Full Tank), kg	44	
Tank Weight, kg	18	
Cost of Gasoline Tank, $	50	
Fuel Consumption, litres/100 km	11.76	
CNG Car Requirements	**Short Driving Range**	**Full Driving Range**
Range, km	340	510
Relative Engine Efficiency	1.1	1.1
Fuel Consumption Increase per 1% Increase in Vehicle Weight	0.60%	0.60%
Cost of Cylinders, $/kg of gas	30-60	30-60
Cylinder Weight, kg/kg of gas	8	8
CNG Car Results		
Weight of CNG to Achieve Full Range, kg	24	38
Extra Vehicle Weight (half-full tanks)	166	280
Extra Vehicle Cost, $	677-1 403	1 080-2 210
Fuel Consumption, litres/100 km gasoline equivalent	11.29	11.70

Note: Prices are in mid-1992 US$.

Table 7.5
CNG Car Additional Cost and Weight
France

Gasoline Car Assumptions		
Range on Full Tank, km	526	
Loaded Vehicle Weight, kg	1 250	
Gasoline Weight (Full Tank), kg	30	
Tank Weight, kg	12	
Cost of Gasoline Tank, $	50	
Fuel Consumption, litres/100 km	7.60	
CNG Car Assumptions	**Short Driving Range**	**Full Driving Range**
Range, km	351	526
Relative Engine Efficiency	1.1	1.1
Fuel Consumption Increase per 1% Increase in Vehicle Weight	0.60%	0.60%
Cost of Cylinders, $/kg of gas	30-60	30-60
Cylinder Weight, kg/kg of gas	8	8
CNG Car Results		
Weight of CNG to Achieve Full Range, kg	16	25
Extra Vehicle Weight (half-full tanks)	110	186
Extra Vehicle Cost, $	433-915	698-1447
Fuel Consumption, litres/100 km gasoline equivalent	7.27	7.52

Note: Prices are in mid-1992 US$.

Maintenance Costs. Differences in the maintenance costs of the cars are due mainly to the differing frequencies of oil and tyre changes. Manufacturers currently recommend more frequent oil changes in diesel cars than in gasoline cars, to compensate for higher rates of oil contamination. CNG cars are generally expected to need less frequent oil changes than gasoline cars. Tyre wear is assumed to be directly proportional to vehicle weight. The distance travelled between oil changes and tyre changes is specified in **Tables 7.2** and **7.3**. The costs of other repairs and replacement parts are also shown. They have been assumed within the spreadsheet to be proportional to the vehicle price and to be incurred every 10 000 km.

Insurance. Insurance premiums are generally correlated with car costs, partly reflecting higher repair costs for more expensive cars, and partly reflecting differences in accident rates. In many instances insurance premiums are determined by engine size, resulting in higher costs for diesel than for gasoline cars. For this analysis, the insurance premium is assumed to have a constant portion, which is half the gasoline vehicle insurance cost, and a variable portion, which is 2.5% of the vehicle cost in France, 2% in the United States.

Fuel Costs. Diesel and unleaded gasoline prices are based on automotive market prices in the United States and France for the second quarter of 1992. CNG prices are calculated using second quarter 1992 gas prices to industry, along with published reports to give a range of filling station costs (DeLuchi *et al.*, 1988; Moreno and Bailey, 1989; USDOE, 1990a). A range of CNG prices has been considered, depending on the costs of filling station operation. **Tables 7.6** and **7.7** give the assumptions used.

Table 7.6
Cost of CNG in United States

Assumptions	Low[1]	High
Compressor Cost, $/TJ/year	1 000	1 750
Compressor Capacity, TJ/year	140	140
Hours of Operation per Year	5 000	3 000
Compressor Housing Cost, $	10 000	20 000
Cost of Land, $	0	50 000
Life of Station Equipment, years	20	20
Compressor Salvage Value (%)	10	10
Wages, $/year	25 000	75 000
Maintenance Costs, $/GJ of CNG	0.25	0.50
Yearly Overheads, fraction of investment (%)	2	2
Electricity Use, kWh/(GJ of CNG)	10	10
Electricity Price ¢/kWh	5	5
Gas Delivered Price, $/GJ	2.65	2.65
Discount Rate (%)	15	15
Results: Costs in $/GJ of CNG Produced		
Electricity	0.50	0.50
Non-Energy Operating Costs	0.60	2.17
Construction Minus Salvage	0.26	0.91
Total CNG Cost	**4.01**	**6.22**

1. Low costs relate to CNG facilities installed in existing gasoline filling stations.

Table 7.7
Cost of CNG in France

Assumptions	Low[1]	High
Compressor Cost $/TJ/year	1 200	2 100
Compressor Capacity, TJ/year	140	140
Hours of Operation per Year	5 000	3 000
Compressor Housing Cost, $	10 000	20 000
Cost of Land, $	0	50 000
Life of Station Equipment, years	20	20
Compressor Salvage Value (%)	10	10
Wages, $/year	20 000	50 000
Maintenance Costs, $/GJ of CNG	0.50	0.25
Yearly overheads, fraction of investment (%)	2	2
Electricity Use, kWh/GJ of CNG	10	10
Electricity Price ¢/kWh	5	5
Gas Delivered Price, $/GJ	3.90	3.90
Discount Rate (%)	15	15
Results: Costs in $/GJ of CNG Produced		
Electricity	0.50	0.50
Non-Energy Operating Costs	0.54	1.67
Construction Minus Salvage	0.31	1.05
Total CNG Cost	**5.25**	**7.11**

1. Low costs relate to CNG facilities installed in existing gasoline filling stations.

Taxation. Fuel tax on CNG is assumed to equal that on gasoline, on an energy-equivalent basis, in both France and the United States. Fuel taxes on gasoline and diesel are assumed to be the same as in the second quarter of 1992. The 1992 Energy Policy Act in the United States provides for temporary tax credits for alternative fuel vehicles, but these have not been taken into account in the results presented in this chapter. The tax credits are expected to make ownership and operation of CNG cars significantly cheaper than gasoline cars, at both the 5% and the 30% discount rate.

Annual Kilometrage and Car Life. The annual distance covered by the car is a key determinant of the economics of different fuels. There is a wide variation in the annual kilometrage driven by each driver in any country. The average distance per year for the first owners of cars is generally higher than that for subsequent owners. Personal travel surveys based on questionnaire responses in the United States and in Europe indicate a rapid reduction in annual kilometrage as cars age. Yet odometer surveys in the United States indicate much less variation, with first owners averaging about 20 000 km/year compared to a national average of about 16 000 km/year (see **Figure 7.2**). All cars have been assumed to have a maximum life of ten years. Maximum lifetime kilometrage is assumed to be 160 000 km in France, 200 000 km in the United States.

In generating a simple cost model for cars, a number of other questions have to be addressed:

- **Discount Rate.** Several rates have been tested, but most of the results presented in this chapter are based on a 5% or 30%. The 5% rate is intended to reflect the opportunity cost of money to buy and operate the car. The 30% rate may be more indicative of choices consumers are likely to make.

Figure 7.2
Variation of Car Use with Age, United States
(kilometres per year obtained from questionnaire and odometer surveys)

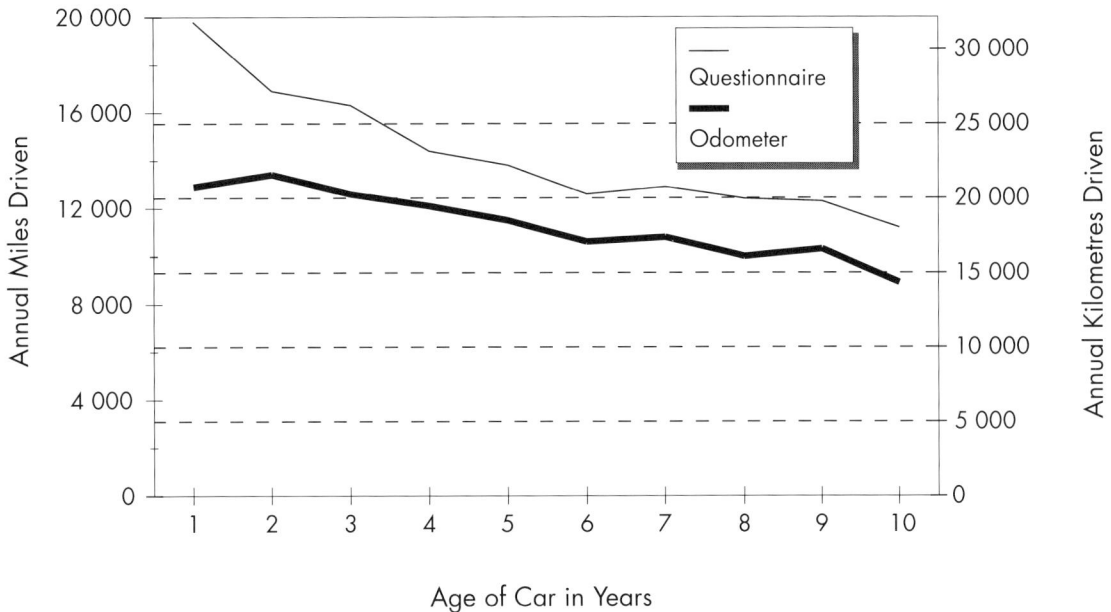

Source: Davis and Morris (1992).

- **Vehicle Life.** Gasoline, diesel and CNG cars wear out at different rates. A well-maintained diesel car can be driven two or more times as far as a gasoline car. CNG engines may also be more durable than gasoline engines, as the fuel and combustion products cause less damage to engine surfaces. Despite these advantages, most cars are scrapped when their bodies cease to be roadworthy, rather than when the engine wears out. In the spreadsheet analysis used here, the lifetime of the car is limited both by the body and the engine. All cars have a maximum life of ten years; the maximum kilometrage can vary between vehicles. Higher kilometrage has been allowed in the United States than in France, reflecting the use of larger, higher specification engines. Diesel vehicles have been allowed 20% higher lifetime kilometrage than gasoline or CNG engines.

- **Vehicle Resale.** Cars generally have at least two successive owners. The first owner might typically keep a car for four years. In France cars are normally sold after this time. In the United States it is becoming more normal for cars to stay about six years in the same household but to be transferred between drivers, for example from parents to their children.

Consumer choices between cars will depend partly on expectations regarding the resale of the cars. Both France and the United States have well-established consumer organisations that provide guidelines for the value of used cars, related to age and kilometrage. Cars are valued roughly in proportion to their expected life — related either to the time before the body deteriorates too far or to the kilometrage before the engine wears out. It is not clear whether used diesel and CNG cars should be valued more highly

107

than used gasoline cars. Indeed, during their first years of use in any market, they are likely to fetch lower prices. The costs presented here, however, all relate to the whole of the vehicle life.

- **Annual Kilometrage.** Calculations have been made over a range of annual travel distances, from 1 000 km to 40 000 km, held constant over the life of the vehicle. In practice cars have high annual kilometrage in their first few years, decreasing to quite low levels just before they are scrapped. In the United States the assumption of constant kilometrage increases the contribution of vehicle purchase and fixed annual costs to the levelised cost, by 1% at a 5% discount rate, and by 16% at a 30% discount rate. In the calculations made for this chapter correction factors have been introduced in the results to allow for this effect. The correction made for France is the same as that for the United States.

7.2   CALCULATION OF LEVELISED COSTS

Costs are calculated as follows:

- The vehicle purchase price is assumed to be paid at time T=0, along with any initial vehicle registration fee and purchase taxes.

- Fixed operating costs — insurance and licence fees — are assumed to be paid at the beginning of each year of vehicle ownership including the first year. They are discounted accordingly and summed.

- Periodic variable operating costs — regular servicing, repairs, replacement of tyres and catalytic converters — are assumed to be paid at fixed kilometrage intervals. The costs are discounted from the time at which they are incurred and summed.

- Fuel costs are treated as effectively continuous. The fuel cost is discounted and integrated over the period of ownership.

- The purchase and fixed operating costs are multiplied by a correction factor to allow for the declining annual kilometrage. This factor is approximately 0.99 at the 5% discount rate, 0.86 at 30%.

- Finally, the distance travelled by the vehicle is also discounted and integrated. The total discounted costs are divided by the discounted distance travelled to give a levelised cost per kilometre travelled.

7.3   RESULTS

Figure 7.3 shows the calculated costs of driving using the various fuels, in France and the United States. These are typical costs over ten years, at a 5% discount rate, of car operation with the national averages for annual distance travelled in 1990. Car purchase dominates the costs in all cases. Taxes are at 1992 levels, except that CNG is assumed to incur the same tax rate as gasoline, on an energy equivalent basis. Switching to either CNG or diesel results in changes amounting to less than 10% in the overall cost of driving.

Figure 7.3
Costs of Driving Gasoline, Diesel and CNG Vehicles, United States and France
(US$/km)

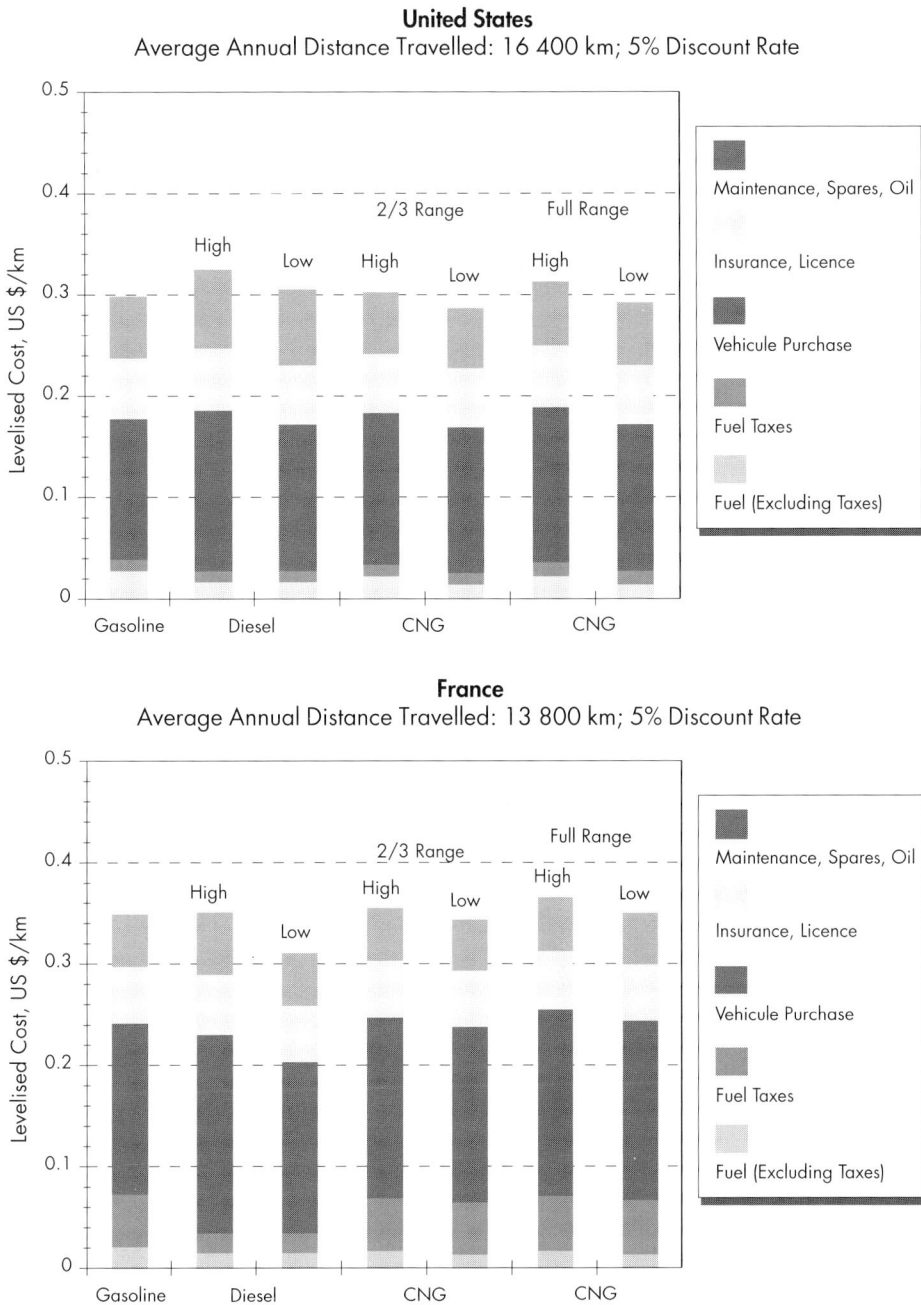

United States
Average Annual Distance Travelled: 16 400 km; 5% Discount Rate

France
Average Annual Distance Travelled: 13 800 km; 5% Discount Rate

The sensitivity of the results to variations in the assumptions has been examined. **Annex 6** contains tables showing the effect on the cost of fuel switching of variations in discount rate, kilometrage, fuel prices and taxes and vehicle efficiency.

Fuel Switching in the United States. Taxes do not have a significant effect on the relative position of different fuels in the United States. The introduction of tax credits mandated by the 1992 Energy Policy Act will change this situation, making CNG more attractive (see **Annex 6**).

Under the conditions examined, the switch from gasoline to diesel is unlikely to be attractive for drivers who travel the average annual distance. For cars with annual use above 24 000 km, the better durability of diesel engines becomes important and they could be cheaper at a 5% discount rate; they are unlikely to be attractive at a 30% discount rate. **Figure 7.4** shows the effect of annual kilometrage on the cost of switching to diesel in the United States, at both discount rates.

Figure 7.4
Cost of Switching from Gasoline to Diesel
United States, 2000
(US cents/km)

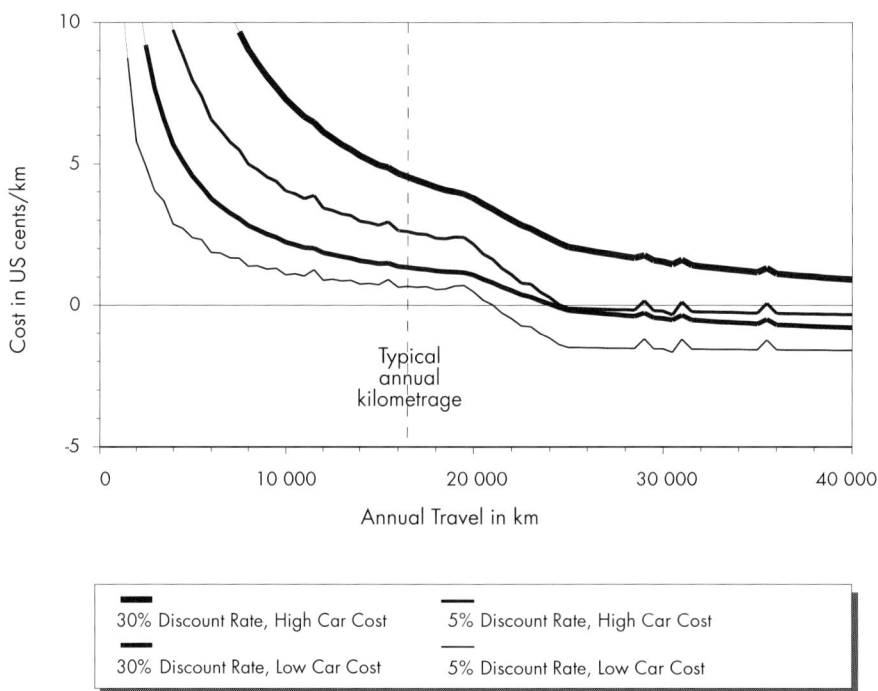

CNG could be cheaper to use than gasoline, provided that drivers are willing to accept a shorter range between refuellings. At a 5% discount rate and without any tax credit, CNG cars are cheaper at average or above average annual kilometrage (see **Figure 7.5**). A discount rate of 30% makes CNG more expensive for most buyers.

If a tax credit is included for filling station operators and is fully reflected in the fuel price, CNG becomes cheaper for the vast majority of car purchasers, at both the 5% and 30% discount rates.

Fuel Switching in France. France's high fuel taxes have a significant effect on the economic choices facing consumers, as **Figure 7.5** shows. Switching to diesel is likely to offer considerable financial savings, as the tax on diesel fuel is 40% lower than that on gasoline. At the lower diesel car price tested, the switch to diesel results in lower costs even without a fuel tax benefit.

Figure 7.5
Cost of Switching from Gasoline to Diesel or CNG
(US cents/km)

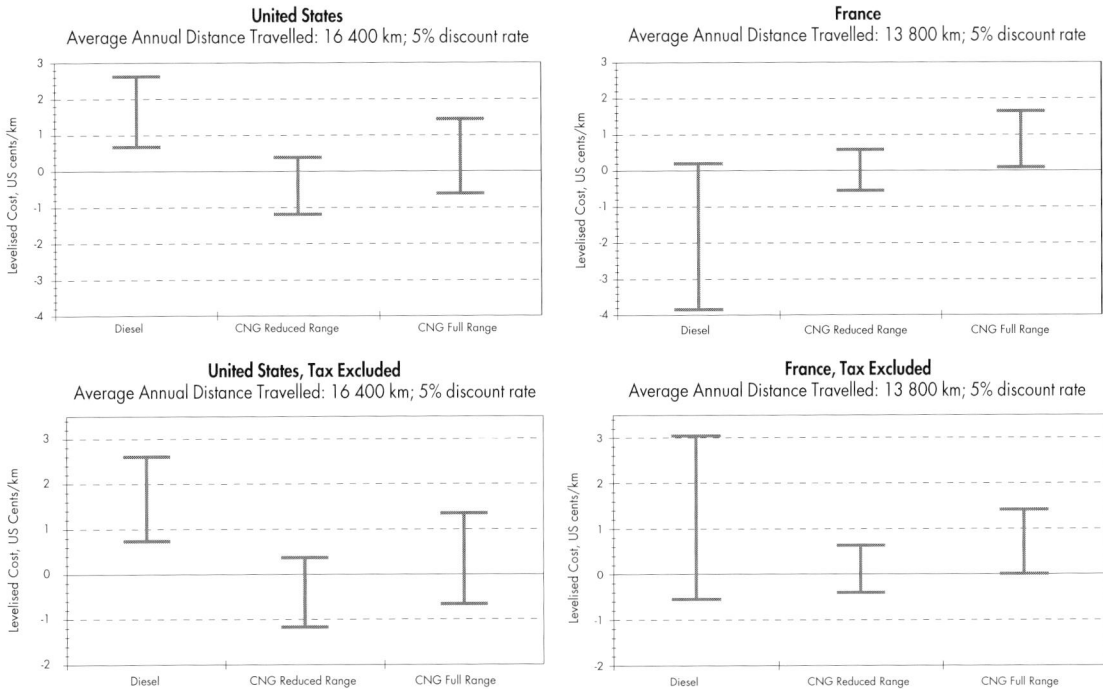

CNG cars with full driving range are likely to be more expensive to use than gasoline, regardless of the discount rate. With or without taxes, the band of costs for CNG cars with reduced driving range straddles the cost for gasoline cars at the 5% discount rate. At a 30% discount rate, reduced-range CNG cars are cheaper only at very high annual travel levels, in the region of 30 000 km.

7.4   BREAKDOWN OF FUEL SWITCHING COSTS

Figure 7.6 shows the contribution to the fuel switching cost made by different elements in the overall cost of car operation. In all cases, fuel costs decrease while costs related to the vehicle increase. Repair costs have been linked in the calculations to car price, but the lower oil change frequency for CNG cars can result in a net decrease in maintenance costs. The high frequency of servicing for diesel vehicles has a significant effect, resulting in an overall increase in variable costs apart from fuel. Fixed costs include insurance, which has been linked to car price, so that there is an increase for switching to diesel or CNG.

111

Figure 7.6
Cost Factors in Fuel Switching, United States and France
(US cents/km)

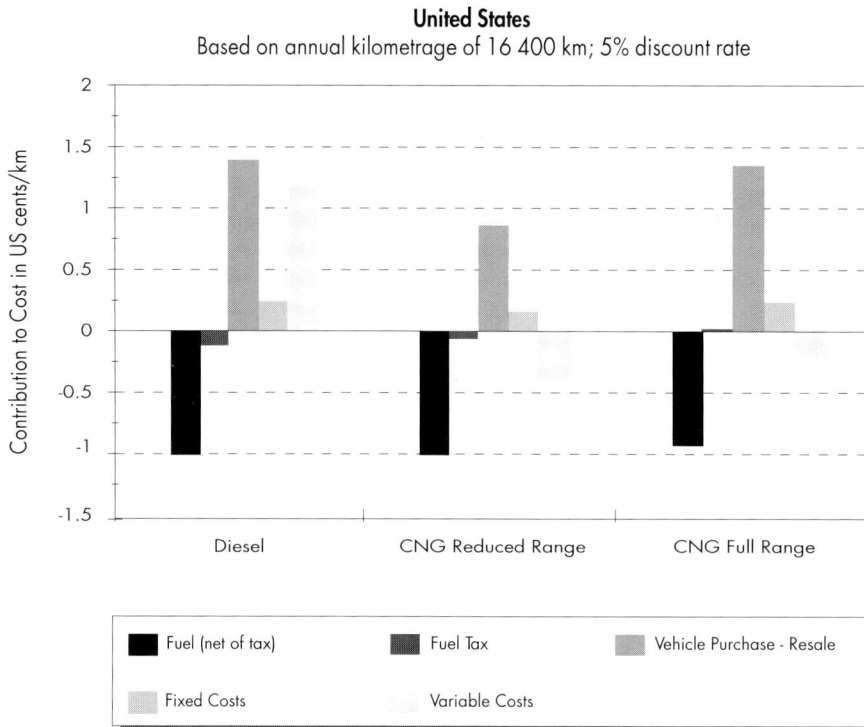

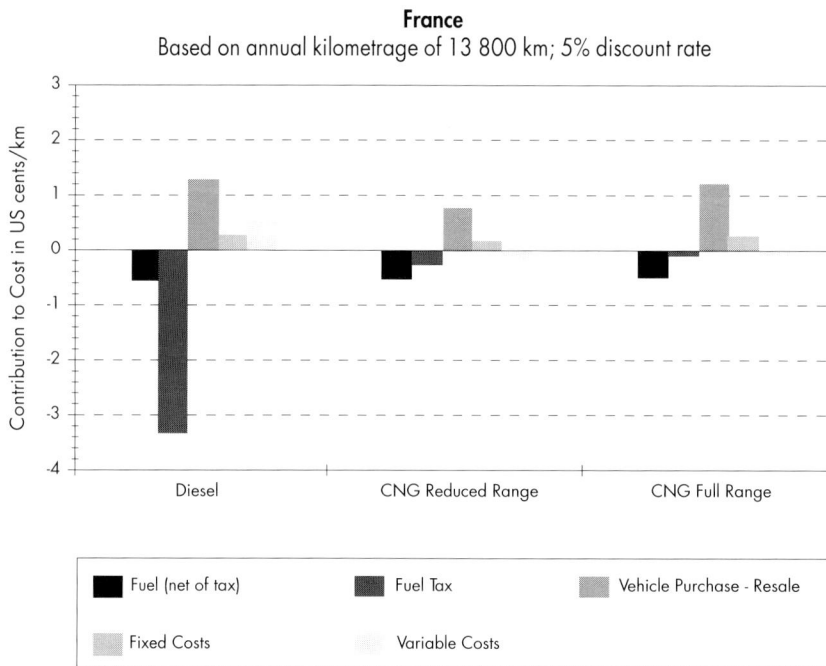

United States
Based on annual kilometrage of 16 400 km; 5% discount rate

Contribution to Cost in US cents/km

Diesel   CNG Reduced Range   CNG Full Range

- Fuel (net of tax)
- Fuel Tax
- Vehicle Purchase - Resale
- Fixed Costs
- Variable Costs

France
Based on annual kilometrage of 13 800 km; 5% discount rate

Contribution to Cost in US cents/km

Diesel   CNG Reduced Range   CNG Full Range

- Fuel (net of tax)
- Fuel Tax
- Vehicle Purchase - Resale
- Fixed Costs
- Variable Costs

Market Niches for Diesel and CNG. Both CNG and diesel cars become more financially attractive for drivers with high annual kilometrage. The switch from gasoline to diesel results in cost savings only where diesel cars are less than 5% more expensive than gasoline cars, where resale values reflect the better durability of diesel engines or where the use of diesel is promoted with tax incentives.

Despite their low overall cost, CNG cars with limited driving range will be of interest only to a small proportion of car users — notably fleet operators whose cars have moderate and predictable daily kilometrage. Their cost per kilometre is affected by annual kilometrage far less than that of diesel cars, as they have been assumed to have the same limit on lifetime kilometrage. The higher cost and bulky fuel storage of full-range CNG cars make it unlikely that they would be taken up without some form of incentive.

7.5   COST-EFFECTIVENESS FOR GREENHOUSE GAS EMISSION REDUCTION

The cost-effectiveness of fuel switching in cars as a means of reducing greenhouse gas emissions has been examined by combining the results of this chapter with those of **Chapter 6**. Costs have been calculated per metric ton of CO_2 equivalent emissions avoided in the whole of the fuel and vehicle life-cycle. This results in lower apparent costs than those obtained by considering only tailpipe CO_2 emissions. The results can be seen in **Figure 7.7**.

Figure 7.7
Greenhouse Gas Abatement Costs, United States and France
(mid-1992 US $/metric ton CO_2 equivalent tax excluded)

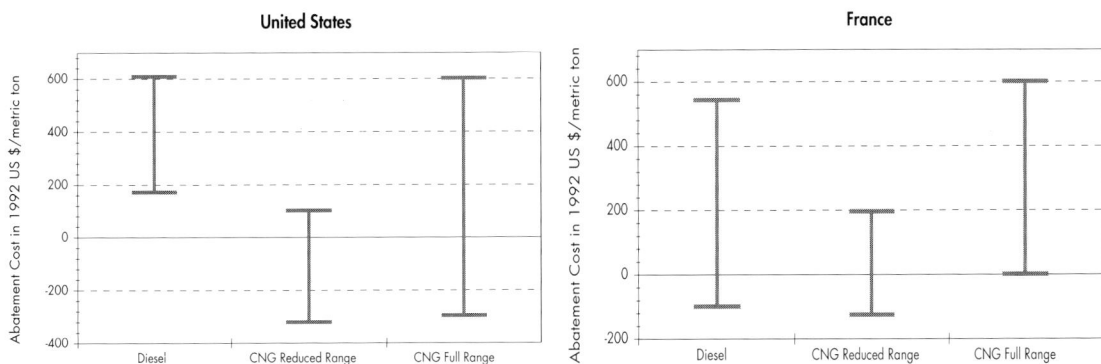

In all cases, emission reduction costs are below $600 per metric ton of CO_2 equivalent and, apart from diesel in the United States, there could be a financial benefit from switching. The use of CNG in cars with reduced range is likely to offer reduction costs below $200 per ton in both France and the United States. The other options could have higher costs than these in both countries.

7.6   POTENTIAL FOR GREENHOUSE GAS EMISSION REDUCTION

While this chapter has identified circumstances in which emissions could be reduced with a financial benefit, the proportion of emissions that could be eliminated in this way is quite limited. The use of CNG was shown in **Chapter 6** to reduce life-cycle emissions from car use by around 10% in 2005. This emission reduction is dependent on the production of cars optimised for CNG, with lower tailpipe methane emissions than are currently achieved. The CNG cars available in the early 1990s may have higher life-cycle greenhouse gas emissions than gasoline cars. Diesel cars offer a greater reduction in life-cycle greenhouse gas emissions, relative to gasoline — in the region of 30%.

7.7   FURTHER WORK

A marginal cost-abatement curve for greenhouse gases cannot be constructed within this report. In order to do so, it would be necessary to have survey information tracking the annual distance driven by cars throughout their lives. From this it would be possible to identify the proportion of vehicle purchasers for whom alternative fuels are cheaper, and the effect of their fuel choice through the life of the vehicle.

A large proportion of the cars bought in both France and the United States are domestically produced, while both countries import oil. The United States also produces most of its own natural gas. Even where fuel switching results in higher costs to the driver, it could boost GDP. Such possibilities require further exploration using economic models.

Econometric studies are needed to obtain a better understanding of the effectiveness of incentives in encouraging drivers to switch fuels. Macroeconomic studies are needed to look at the effects on the whole economy of using incentives to encourage fuel switching.

CHAPTER 8[1]

HEAVY-DUTY DIESEL VEHICLES

Heavy-duty vehicles include goods vehicles and buses over 3.5 metric tons in Europe and 8 500 pounds (3.9 metric tons) in the United States. In Europe the vast majority of these vehicles are diesel powered. In the United States about 82% of fuel use by medium- and heavy-duty trucks is diesel (Davis and Morris, 1992). Most heavy-duty vehicles have direct-injection (DI) diesel engines, in contrast with most light-duty vehicles in Europe, which use indirect injection (IDI), or swirl chamber, engines. As DI's dominance in heavy-duty diesel engines is likely to continue, this chapter concentrates on such vehicles.

8.1 ENERGY USE IN HEAVY-DUTY VEHICLES

Most heavy-duty vehicles are used for commercial purposes, and their operators place a high priority on minimising costs. Fuel use can be a significant cost element, ranging typically from about 10% in smaller trucks and buses to 20% or more in large trucks. Wages account for most of the rest, varying significantly between countries.

The variety in design of heavy-duty vehicles is much greater than in light-duty vehicles, and it is not particularly helpful to present energy use or emissions on a vehicle-kilometre basis. **Chapter 3** presents some energy consumption data per passenger- and ton-kilometre, which is useful for comparing modes and vehicle types. This chapter focuses on vehicle technology. For a given vehicle class, energy use and emissions per unit of payload are affected by several factors:

- **Load Factor.** Most commercial operators try to maximise vehicle loading and hence revenue. Bus load factors have generally fallen over the last two decades (Davis and Morris, 1992). In some truck operations, especially for large, contract haulage trucks, load factors are improving. There is increased use of advanced information technology to optimise fleet use. With increasing loads, vehicle energy use increases, but energy use per unit of load decreases. Practices such as "just-in-time" manufacturing tend to result in higher vehicle kilometrage with lower load factors.

---

1. Based on material supplied by Ricardo Consulting Engineers Ltd., Shoreham-by-Sea, UK and Innas, Breda, the Netherlands.

- **Driving Pattern.** Vehicle energy use and emissions are high in urban traffic and at high speeds. Energy use can vary by 10-20% according to the skill of the driver.

- **Engine Maintenance.** Modern, well-maintained diesel engines should not produce visible particulate emissions. Poor maintenance, especially of injection pumps and nozzles, can result in high emissions at full load and reduced energy efficiency due to slow, incomplete fuel combustion.

- **Tyre Design and Air Pressure.** Tyre design is becoming increasingly sophisticated. The significance of rolling resistance in truck energy use is increasing as air resistance decreases with improved aerodynamic design. Rolling resistance can be reduced by replacing twin tyres with double-width "super singles" and by using improved materials, low-profile tyres and steerable rear axles. Reduced tyre pressure also increases rolling resistance.

- **Engine and Transmission Design and Size.** Energy use increases with the ratio of engine power to vehicle weight, though the effect is more significant in gasoline vehicles. It is also important to match the transmission to the engine, the vehicle and the application. Urban buses generally have automatic transmissions, unlike very large trucks that do a lot of highway driving. There is a wide variety of automatic transmission types for heavy-duty vehicles. Many systems have a driver-operated clutch with a servo-operated gear change and some form of indicator to let the driver know when to change gear. Integrated engine/transmission management systems are being developed.

- **Body Design.** Considerable potential remains for reducing drag, especially in large trucks. The desirability of minimising body production costs and maximising load capacity, within height and width limits, has tended to result in box shapes. Rounding corners, smoothing vehicle surfaces and fairing-in the gaps behind the cabs can halve air resistance. There is less immediate potential for reducing weight, as manufacturers have already paid considerable attention to this in order to maximise payload capacity.

One estimate of the potential for energy efficiency improvements is 10% by 2010, through changes in body design and transmissions (Martin and Shock, 1989). In the same period, engine efficiency improvements could result in an additional 4% energy savings. The introduction of tighter emission controls may result in lower engine efficiency, however, and there could be a net worsening of fuel economy. The rest of this chapter focuses on the potential for engine energy efficiency improvements, emission reduction and the use of alternative fuels.

8.2   CURRENT ENGINES

The DI engines that power most medium- and heavy-duty trucks and buses in North America and Europe offer a good compromise in terms of energy efficiency, power/weight ratio, durability and cost.

VOC and carbon monoxide emissions are relatively low and do not generally cause concern. The diesel engine emissions that attract the most attention are NO_x and particulates. These emissions are addressed in standards due to come into force in the mid-1990s. **Figure 8.1** shows the trade-off between NO_x and particulates for typical engines tested on the UN ECE and US federal test cycles. The graphs also show the emission standards for the European Community and the United States.

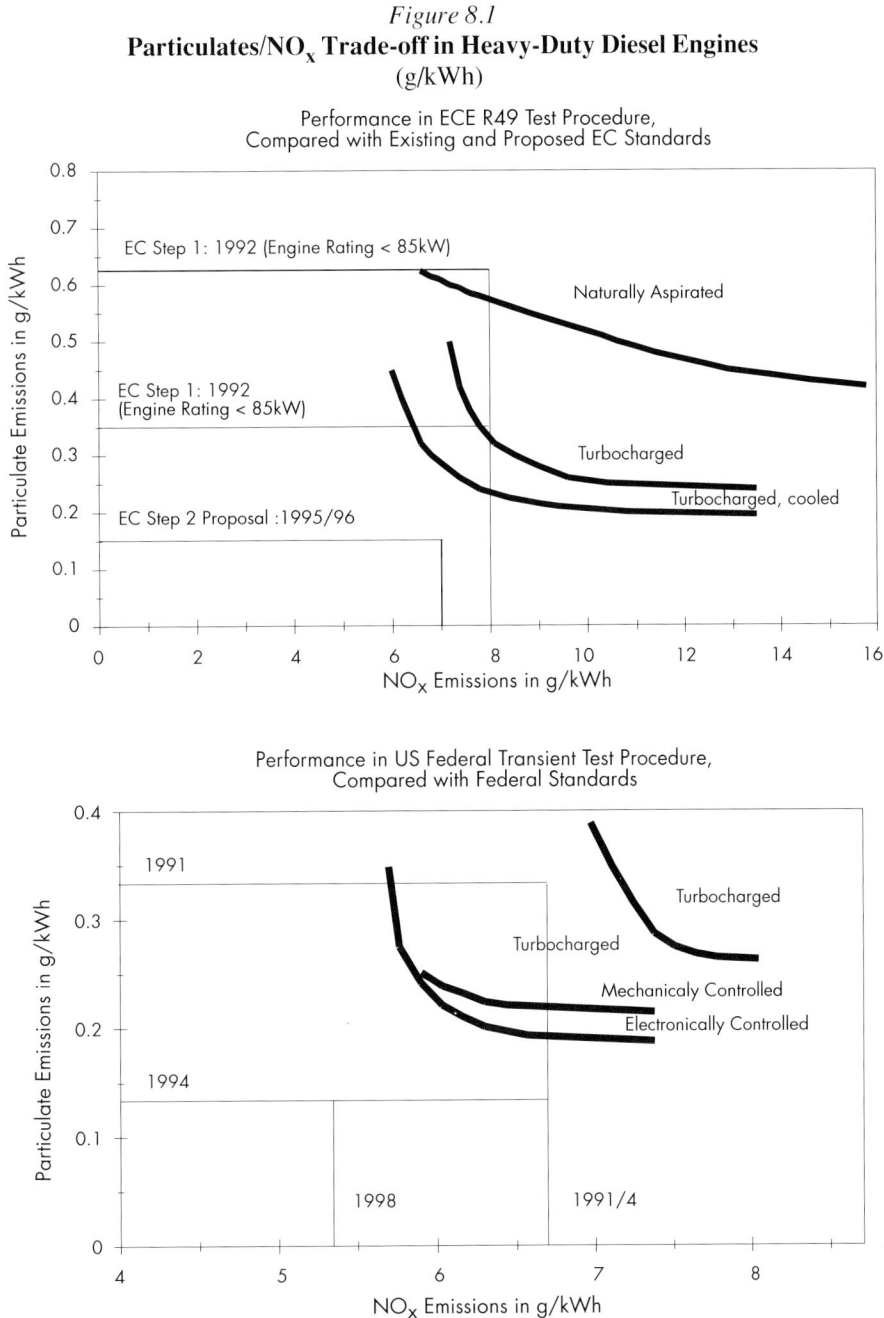

Figure 8.1
Particulates/NO_x Trade-off in Heavy-Duty Diesel Engines
(g/kWh)

Performance in ECE R49 Test Procedure,
Compared with Existing and Proposed EC Standards

Performance in US Federal Transient Test Procedure,
Compared with Federal Standards

Note: Curves indicate the lowest emissions obtained with each engine type.

Source: Ricardo (1992).

117

Simultaneous control of particulates and NO_x is difficult. In the past manufacturers strove to reduce particulates and fuel consumption through optimum fuel injection timing. Some reduction in NO_x can be achieved by retarding injection, at the expense of higher particulates and fuel consumption. While existing engines can meet current standards in this fashion, further engine development will be needed to meet some of the tighter standards planned.

8.3   LEGISLATED EMISSION REQUIREMENTS

European and US engine manufacturers offer a range of engines, including older models, developed to their maximum potential to meet current emission requirements. They also offer new models designed specifically for low emissions. The newer models will be adaptable to standards in 2005 and beyond.

Heavy-duty engines currently being certified in all major markets are of conventional, if advanced, technology. The main visible changes in recent years are in the use of turbochargers, charge cooling and, in a growing number of models, electronically controlled fuel injection systems.

US particulate emission limits effectively rule out highway use of the naturally aspirated engines currently available. EC rules allow a higher particulate level for engines of less than 85 kW, permitting the use of naturally aspirated engines in large panel vans and light trucks. Even in these vehicles there is greater use of turbocharging. EC standards for 1995 will probably lead to discontinued use of naturally aspirated engines for highway applications.

Legislation to reduce exhaust emissions is being tightened throughout the OECD. Current limits have already required attention to fuels and engine technology, while future limits will stretch requirements for both fuels and engines beyond existing technology.

United States. Limits on gaseous pollutants from heavy-duty engines were introduced in the United States in 1979. By 1983 the current transient test procedure had been adopted. The Federal Test Procedure (FTP) cycle is intended to be representative of US driving conditions (see **Figure 8.2**). It is biased heavily towards high speeds and light loads. The full torque potential of the engine is barely achieved during the test cycle.

US emission standards for heavy duty engines are shown in **Table 8.1**. The standards were implemented from 1991 and the engine manufacturing industry is now concentrating on meeting 1994 standards. The US market has readily accepted turbocharging and charge cooling as the medium-term route to low emissions.

The California Air Resources Board (CARB) has presented plans for heavy-duty vehicles analogous to the state's legislation for light-duty vehicles, discussed in **Chapter 9** (CARB, 1992). The proposal is shown in **Table 8.2**. Low-emission and ultra-low emission trucks would each have captured 50% of the market by 2004. These proposed emission standards are significantly more stringent than the federal levels proposed for the same period.

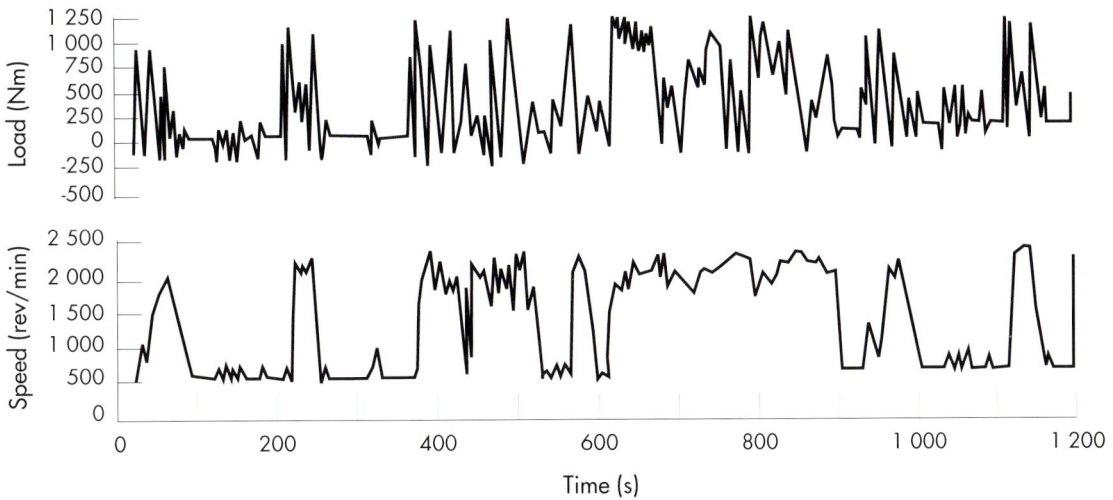

Figure 8.2
The US Federal Transient Test Cycle

Source: Innas (1992).

Table 8.1
US Emission Legislation, Heavy-Duty Vehicles
(g/hp h)

Levels in 13-mode cycle

	VOC+NO$_x$	Carbon monoxide	VOC	
Late 1970s	10	25	1.5	

Levels in FTP transient cycle

	VOC	Carbon monoxide	NO$_x$	Particulates
1983	1.3	15.5	10.7	0.6
1991	1.3	15.5	5.0	0.25
1993 (Buses)	1.3	15.5	5.0	0.25
1994	1.3	15.5	5.0	0.1
1994 (Buses)	1.3	15.5	5.0	0.05[1]
1998	1.3	15.5	4.0	0.1 (Federal proposal)
2004	1.3	15.5	3.0	0.1 (California proposal)

1. Or 0.07 if 0.05 g/hp h proves infeasible.

119

Table 8.2
CARB Proposals for Heavy-Duty Engine Standards
(g/hp h)

Trucks

	NO_x	Particulates	THC[1]	RANMOG[2]	Carbon monoxide	Formaldehyde	Aldehydes
1994 Base	5.0	0.1	1.3	1.2	15.5	0.05	0.09
1998 Base	4.0	0.1	1.3	1.2	14.4	0.05	0.09
2004 Base	3.0	0.1	1.3	0.9	14.4	0.05	0.09
TLET[3] (Introduction 1995)	3.0	0.1	1.3	0.9	14.4	0.05	0.09
LET[4] (Introduction 1996)	2.0	0.08	1.3	0.6	14.4	0.05	0.09
ULET[5] (Introduction 1999)	1.0	0.05	1.3	0.3	7.2	0.05	0.09
ZET[6] (Date undecided)	0	0	0	0	0	0	0

Buses

	NO_x	Particulates	THC[1]	RANMOG[2]	Carbon monoxide	Formaldehyde	Aldehydes
1994 Base	5.0	0.05	1.3	1.2	15.5	0.05	N.A
1998 Base	2.5	0.05	1.3	1.2	15.5	0.05	N.A
LEB[7] (Introduction 1999)	2.0	0.05	1.3	0.6	14.4	0.05	0.09
ULEB[8] (Introduction 1999)	1.0	0.05	1.3	0.3	7.2	0.05	0.09
ZEB[9] (Introduction 2005)	0	0	0	0	0	0	0

1   Total hydrocarbons as measured with a flame ionisation detector.
2   Reactivity adjusted non-methane organic gases — an indicator of exhaust gases' ozone formation potential.
3   Transitional low-emission trucks.
4   Low-emission trucks.
5   Ultra-low-emission trucks.
6   Zero-emission trucks.
7   Low-emission buses.
8   Ultra-low-emission buses.
9   Zero-emission buses.

Europe. Emissions from heavy-duty engines in the EC are controlled according to Directive 91/542[1], as shown in **Table 8.3**. From 1992 limits have been placed on particulate matter, with a further tightening of the standards, known as Stage 2, scheduled for 1995. Discussions are under way on Stage 3, which is expected to be introduced in 1999.

Table 8.3
Emission Regulations in Europe, Heavy-Duty Vehicles
(g/kWh)

Levels in ECE-R49 (13-mode)

	VOC	Carbon monoxide	NO$_x$	Particulates
Introduction	3.5	14.0	18.0	
1992	1.1	4.5	8.0	0.36-0.61[1]
1995	1.1	4.0	7.0	0.15

1. Depending on engine rating.

The engine test cycle used for certification is a 13-mode, steady-state procedure, illustrated in **Figure 8.3**. The weighting factors applied to the 13 modes reflect patterns of usage for heavy-duty vehicles in Europe. The test procedure is heavily biased to operation at peak torque. The potential for incorporating an element of transient operation into the R49 test procedure is being discussed.

Figure 8.3
The European 13 Mode Test Cycle

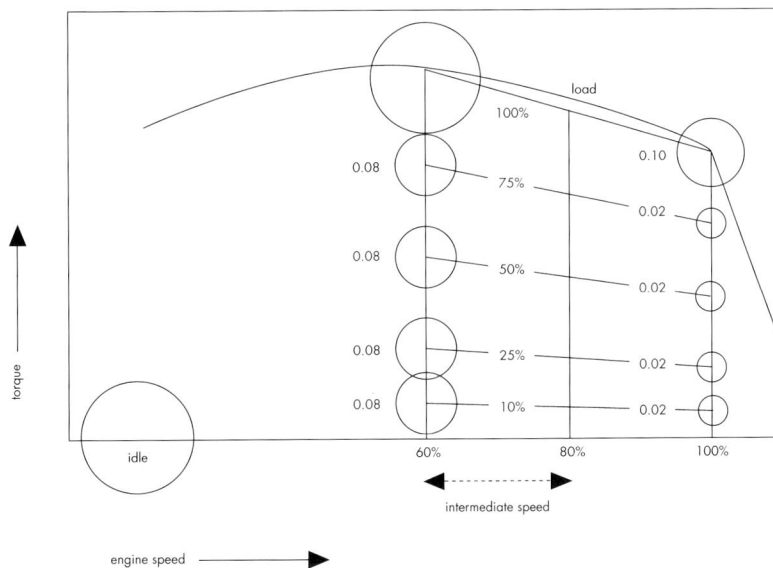

Note: Circle size represents the weight factor for each mode.
Source: Innas (1992).

1. This directive amended directive 88/77, adopted from ECE Regulation 49.

Countries of the Stockholm Group[1] have also adopted the European 13-mode test procedure. Standards and introduction dates differ between countries.

8.4   IMPROVEMENTS IN ENGINE TECHNOLOGY

The basic operation of a light-duty diesel engine is briefly described in **Chapter 4**. Heavy-duty diesel engines work essentially the same way. This section examines opportunities for emission abatement through modifications to such engines. The opportunities are related to the best commercial technology, giving 5.5 g/kWh of NO_x and 0.13 g/kWh of particulates in the FTP transient cycle.

Table 8.4 summarises the performance of a range of technologies that could be incorporated into engines in the next few decades. Though some are not expected to be commercial before 2005, considerable research effort has been devoted to them and theoretically they can bring substantial energy efficiency improvements. Most of the technologies are of interest because of their potential to reduce NO_x or particulate emissions, especially where this can be done without sacrificing energy efficiency.

Air Charge Preparation. Turbocharging and aftercooling are likely to be standard on all heavy-duty engines by 2005. They offer an inexpensive way to simultaneously improve power/weight ratios, fuel economy and levels of NO_x and particulate emissions.

Exhaust gas recirculation (EGR) is considered one of the primary NO_x control technologies for the future. Extensive development is needed with regard to combustion system design, the recirculation and admission system, cooling at high load, rematching of turbomachinery, wear limitation and transient control.

Fuel Injection Equipment. To achieve NO_x emissions of about 2.75 g/kWh in the FTP transient cycle, several injection systems are possible. Some are in production now. They include:

- advanced rotary pump with electronic control, appropriate only for medium-duty engines;

- electronic unit injector, the technology now most favoured by manufacturers. It is likely to be the dominant system for heavy-duty and some medium-duty engines by 2005. It can deliver injection pressure of over 1 700 bar, though this might be too high for very low NO_x requirements (Frankl *et al.*, 1988);

- electronic sleeve control in-line fuel injection equipment, suited to medium- and heavy-duty engines. Potential injection pressure is about 1 500 bar with development. Performance is comparable to that of electronic unit injectors (Nishizawa *et al.*, 1987; Stumpp *et al.*, 1989);

- common rail/accumulator system, a pre-production solution offering potentially the greatest flexibility of injection rate control and timing. This system is ideal for use with EGR. The cost is high, possibly twice that of existing fuel injection equipment. More work is needed (Miyaki *et al.*, 1991; Beck *et al.*, 1984).

---

1.   Austria, Denmark, Finland, Norway, Sweden and Switzerland.

Table 8.4
New Technology for Heavy-Duty Diesel Engines
(relative to current best commercial technology[1])

Technology	Effect on Emission Levels NO$_x$	Particulates	Percentage BSFC Decrease	Development Status	Other Comments
Air Preparation/ Combustion Chamber Design					
EGR	down 50-90%	up	2-3	up to 5 years	
Turbocompound	down	no change	1-3	commercial	large trucks only
Charge cooling	down 10%	not known	not known	up to 5 years	
Better injector design in low-swirl engine	no change	20%	no change	by 1996	large engines
Better combustion chamber in high-swirl engine	down 10%	down 20%	not known	by 1996	medium size engines
Injector nozzle placement in high-swirl engine	no change	down 5%	no change	by 1996	
Compression ratio	not likely to change				
Ceramic engine lining	up 50%	up	up	10-20 years	
Water injection	down 20%	up 5%	no change	5-10 years	
Fuel Injection System					
Advanced rotary pump	no change	no change	no change	5 years	small trucks only
Sleeve control	no change	no change	no change	commercial	
Common rail	no change	no change	no change	5-10 years	
Lubrication					
Oil formulation	minimal changes possible				
Engine manufacturing tolerances	gradual progress expected; will reduce variation between engines				
Friction					
General improvements	2%	2%	2%	5 years	
Exhaust Treatment					
Particulate traps	up 10%	down 75%	up 10%		Traps commercial but regeneration requires more research
Lean NO$_x$ catalyst	down 20-50%	no change	no change	10-15 years	
Oxidation catalyst	no change	down 20%	no change	commercial	

1   Turbocharged, aftercooled, electronic unit injection engine, with rated power of 200kW. Emissions in FTP: 5.5g/kWh of NO$_x$, 0.13g/kWh of particulates; brake specific fuel consumption (BSFC) of 190-220g/kWh.

8.5   EXHAUST TREATMENT

Particulate Traps. Traps are the most effective way to reduce particulate emissions, with over 75% removal possible. Trapping particulates is relatively easy and a large variety of materials and approaches is available. Removing the trapped soot is a major problem, however, despite considerable effort to develop trap regeneration systems.

At this stage traps can be used in city buses but are not a practical solution for heavy goods vehicles. Existing systems can be more expensive than the engine itself. Cheaper traps may become available by the late 1990s. They will still have an adverse effect on fuel economy, because of the increase in exhaust back pressure imposed on the engine and the energy required for regeneration.

Lean NO$_x$ Catalyst. The diesel engine operates with lean air/fuel mixtures, so three-way catalytic converters are not feasible. Some research has been done using ammonia or urea to reduce NO$_x$ but such methods are not practical for road vehicles (Held *et al.*, 1990).

Catalyst and engine manufacturers are engaged in a large-scale research effort on NO$_x$ reduction by zeolite catalysts. To date, about a 20% reduction is being achieved (Schoubye *et al.*, 1987). Reactive hydrocarbons in large quantities can increase NO$_x$ reduction to about 50%. The catalyst is inhibited by water in the exhaust and poisoned by sulphur from the fuel.

Catalytic converter manufacturers are optimistic that the lean NO$_x$ catalyst will be available by the end of the 1990s. If a catalytic converter with 30% efficiency were to be available by then, it would enable low NO$_x$ levels to be achieved without recourse to retarded timing, and hence good fuel consumption could be maintained.

Oxidation Catalysts. Oxidation catalysts are a mature technology for light-duty applications and have been demonstrated to be effective for heavy-duty engines in the FTP (Porter *et al.*, 1991). They have very little effect on NO$_x$ emissions, but can reduce VOCs and carbon monoxide by up to 80%. They also reduce odour-related compounds such as aldehydes. The catalyst can oxidise most of the soluble organic compounds that make up about 20% of particulates.

The main difficulty with the use of oxidation catalytic converters on heavy-duty diesel engines is that they can cause the formation of sulphuric acid and sulphates from sulphur dioxide in the exhaust. These make a considerable contribution to particulate mass (see **Figure 8.4**). Current fuel sulphur levels typically range from 0.1% to 0.4% by mass; levels as low as 0.05% can lead to significant sulphate formation at high load.

The durability of oxidation catalytic converters on heavy-duty engines has yet to be determined, but it is expected to be acceptable. They have a negligible effect on fuel consumption.

Estimated emission requirements in 2005 and their implications for fuel consumption are summarised in **Table 8.5**. For this table it is assumed that EGR or lean NO$_x$ catalysts become available by 2005. EGR is under development and results so far are promising. Nevertheless, it is not certain that the durability problems can be solved. Lean NO$_x$ catalysts are a decade away from widespread use. If neither option is available by 2005, the penalty in fuel consumption will be nearer to 10-12%, as greatly retarded injection timings will then have to be adopted. Diesel vehicles would not be usable in California in that case and would need to be replaced by methanol, CNG or gasoline engines.

Figure 8.4

Effect of Oxidation Catalyst on Composition of Particulates from Heavy-Duty Diesel Engines
(g/kWh)

European 13 - Mode Test

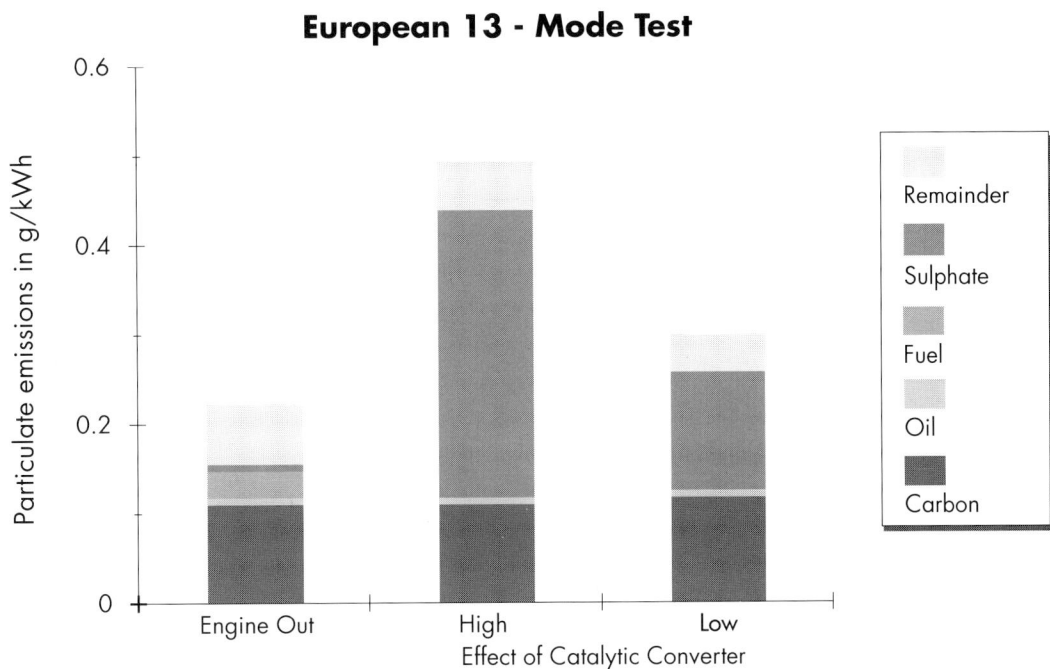

US Federal Transient Test

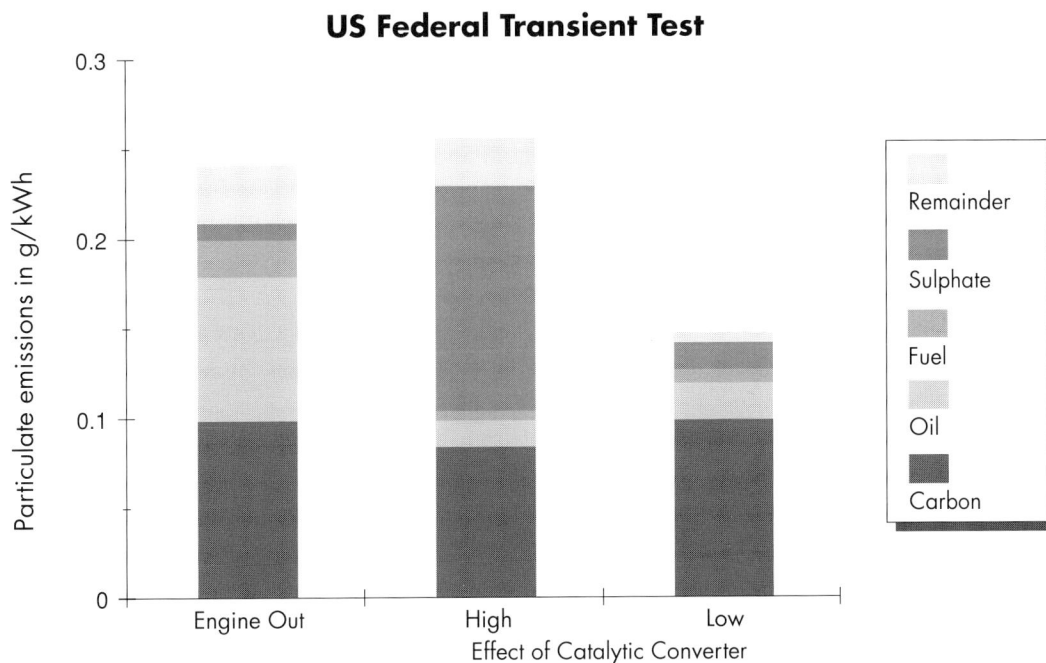

Source: Ricardo (1992).

Table 8.5
Technology to Statisfy 2005 Emission Regulations for Heavy-Duty Vehicles

| | Regulations | | Technology | BSFC Change[1] (%) |
	NOx	Particulates		
US Federal[2]	3g/hp h	0.1g/hp h	Diesel+EGR+Electronic Injection, or	+ 5
			Diesel+ DeNO$_x$ Catalyst	0
California[2]	2g/hp h	0.05g/hp h	Methanol/CNG, or	+20
			Diesel+EGR+Trap, or	+10
			Diesel+DeNO$_x$+Trap	+ 7
Europe[3]	5g/kWh	0.1g/kWh	Diesel+EGR, or	+ 5
			Diesel+DeNO$_x$	0

1   Mass-based BSFC expected for 2005 engine, relative to 1992.
2   Transient cycle.
3   R49 cycle (transient elements may be added).

8.6   FUEL QUALITY

Fuel Changes Necessitated by Engine Technology. Existing engines are quite tolerant of variations in fuel quality. If, as seems likely, oxidation catalytic converters are widely introduced, the tendency towards higher sulphate formation could become a problem. With current 0.05% sulphur fuels and an engine meeting mid-1990s standards without exhaust treatment, about 10% of the resulting particulate emissions are likely to be sulphates and associated bound water. A further 20% will be soluble organic compounds. Catalytic reduction of soluble organics could lead to much higher sulphate levels unless fuels and catalytic converters were carefully formulated.

Catalytic converters need further development to reduce conversion of sulphur dioxide to sulphates. Further reductions in fuel sulphur, ideally to below 0.01%, are also needed.

Fuel Changes Aimed Specifically at Reducing Emissions. Diesel fuel modifications offer a way to reduce exhaust emissions from engines not controlled by catalytic converters.

Sulphur. Fuel certification levels have been set at 0.1% for US 1991 and Europe 1995 standards, and 0.05% for US 1994 and Europe 1996 standards. At 0.05%, fuel sulphur contributes about 10% of the particulate emissions of a US 1994 specification engine operating within the required 0.13 g/kWh in the FTP.

Sweden has the world's tightest limits on sulphur in diesel fuel, with tax incentives for the lowest-sulphur fuels. Three tax classes of fuel are defined — Class III: 0.2% sulphur, Class II: 0.02% sulphur, and Class I: 0.001% sulphur. Aromatics and volatility are also defined.

Cetane Number. European diesel fuel has cetane numbers typically around 48-52. In the United States, grade #1 diesel oil has cetane numbers of 40-44 and grade #2 is about 32-37. Work in the United States demonstrates that high-cetane fuel brings several advantages: improved combustion; reduced emissions of VOCs, carbon monoxide and particulates; improvements in cold starting and reductions in white smoke after starting; and less noise (Lange, 1991).

126

Engine manufacturers' associations in the United States and Europe are demanding higher cetane numbers. In Europe, manufacturers have called for cetane numbers of 58 or more. It is unlikely that the oil industry will be able, or required, to provide such fuel for general use. The United States will probably increase cetane numbers, but no increases are likely in Europe. There is little evidence that emissions improve significantly with fuel cetane numbers above current European levels.

Aromatic Content. Since aromatics are generally found to contribute to soot formation, low aromatic content will probably be a necessary feature of "green" diesel fuel. Reducing aromatic content can cut particulate emissions by up to 14%, in addition to a small NO_x reduction (Lange, 1991; Bennethum and Windsor, 1991).

Density. A decrease in fuel density from the typical European level of 845 kg/cubic metre to 820 kg/cubic metre could reduce particulate emissions by up to 8%. "Low emission" fuels of the type used in Sweden have density of about 800 kg/cubic metre. If the fuel delivery is reset to give constant mass fuelling, the case for density having effect on emissions becomes uncertain.

Volatility. Lowering a fuel's boiling point can reduce particulate emissions (Weldmann *et al.*, 1988). There is also evidence of effects from variation in initial boiling point, but any benefits are small compared to those gained by other factors, and emission variations due to volatility alone are not well reported.

Reformulated Diesel. Reformulated fuels are being introduced in some urban areas for vehicles such as buses and garbage trucks. Because of their cost, however, they are unlikely to be extended rapidly to general use. Reducing the amount of aromatics or heavier hydrocarbons in diesel fuel will mean a lower diesel yield from crude oil, or tighter conversion processing requirements. Diesel fuel desulphurisation requires refinery investment in new plant, and in itself uses energy. Some reformulated fuels are likely to have higher life-cycle CO_2 emissions than current diesel fuels.

8.7   ALTERNATIVE FUELS

Methanol. Significant effort has been made to develop methanol and ethanol engines for heavy-duty vehicles. Methanol can be used in heavy-duty diesel engines, reducing emissions of particulates and NO_x but giving high formaldehyde emissions. Methanol has a very low cetane number, however, resulting in delayed ignition, which leads to high emissions of carbon monoxide and VOCs.

Methanol's energy density is about 44% that of diesel fuel, so fuelling volumes are about 2.25 times those for diesel. Methanol has been the subject of many diesel conversion and evaluation programmes, the best reported being those involving US urban bus and truck fleets. Most programmes report energy efficiency (usually as diesel-equivalent fuel economy) 5-10% worse than for diesel-powered fleets.

Methanol is corrosive and has lower lubricity than diesel fuel. The fuel system requires modifications for methanol compatibility and increased volume. Methanol is unlikely to replace diesel fuel in its main markets in the medium term. It may be used in specialised ultra-low-emission applications, such as urban buses and trucks.

CNG. CNG does not ignite in a compression ignition engine but requires a separate source of ignition energy. Nevertheless there has been considerable interest in using CNG in heavy-duty engines to reduce emissions. Possible approaches are:

- stoichiometric combustion in a spark ignition engine with a three-way catalytic converter, now used in light-duty engines. In heavy-duty engines it can result in very low emissions but energy consumption is 20-25% higher than in DI engines;

- lean combustion in a spark ignition engine with an oxidising catalyst, probably the best approach for heavy-duty engines;

- diesel fuel pilot injection in a compression ignition engine; energy efficiency is similar to that of a diesel engine but emissions, especially of methane, are high (see **Table 8.6**);

- direct injection, high compression-ratio engine with a glow plug, now used only in engines with cylinder bore exceeding 200mm. Energy efficiency is similar to that of a diesel engine, but again methane emissions may be high.

Both of these spark ignition approaches require inlet air throttling to control the air/fuel ratio. Diesel engines can be converted to operate as natural gas spark ignition engines without changing the compression ratio.

Natural gas can be stored in compressed or liquid form, but the high storage volumes and heavy fuel tanks needed make it unattractive for many conventional vehicles. It remains a serious contender in some applications, such as city buses, urban delivery vans and waste collection vehicles in areas where emissions are tightly controlled.

Vegetable Oil Derivatives. Vegetable oils that have been evaluated as diesel replacements include sunflower seed, rapeseed, hemp, soya, coconut and palm oil. All are more expensive than diesel oil, and supplies are limited. Diesel engines run well on such oil for short periods, but serious fouling of the fuel injection equipment soon results in loss of performance.

To overcome this problem, the oil can be heated with alcohol in the presence of a catalyst, converting the triglycerides into less viscous methyl or ethyl esters. The most credible contender is rapeseed dimethyl ester (RME), discussed further in **Chapter 10**. Its VOC emission levels have been reported to be 20% lower than with diesel fuel, carbon monoxide emissions 13% lower and particulates 16% lower. NO_x emissions, however, increase by 20% (Tritthart and Zelenka, 1990). The fuel gives energy efficiency about the same as that of diesel fuel in a DI engine. It is relatively slow burning, however, so for the same energy input the engine produces about 9% less power (Needham *et al.*, 1989).

Greenhouse Gas Abatement Potential. The potential for greenhouse gas abatement using alternative fuels has been examined using the spreadsheet programme applied in **Chapter 6** to light-duty vehicles. **Figure 8.5** shows typical results. It is clear that most alternative fuel options for heavy-duty vehicles do not offer any potential for greenhouse gas abatement. Those that do — wood-derived biofuels and hydrogen — are not likely to be commercially available before 2005. **Table 8.6** summarises the expected effects of alternative fuels on emissions and performance, and the development status of engines using these fuels.

Table 8.6
Alternative Fuel Technology for Heavy-Duty Diesel Engines

Fuel/Technology	NO$_x$	Emissions in g/kWh Particulates	THC[1]	Fuel Consumption	Development Status
CNG, Diesel Pilot Compression Ignition	7-11	0.5-1.0	30-50	no change	commercial
CNG, Stoichiometric with Closed loop λ Control, Spark Ignition	1.6-4.5	~0.01	1.4-1.9	~22% higher	commercial
Methanol	3-10	0.05-2.7	0.5-60	5% lower	prototype
RME		———————— same as diesel ————————			commercial

1   Predominantly methane in CNG engines.

Sources: Sypher (1992); CEC (1992); California Energy Commission (1987).

Figure 8.5
Life-cycle Greenhouse Gas Emissions from Heavy-Duty Vehicles

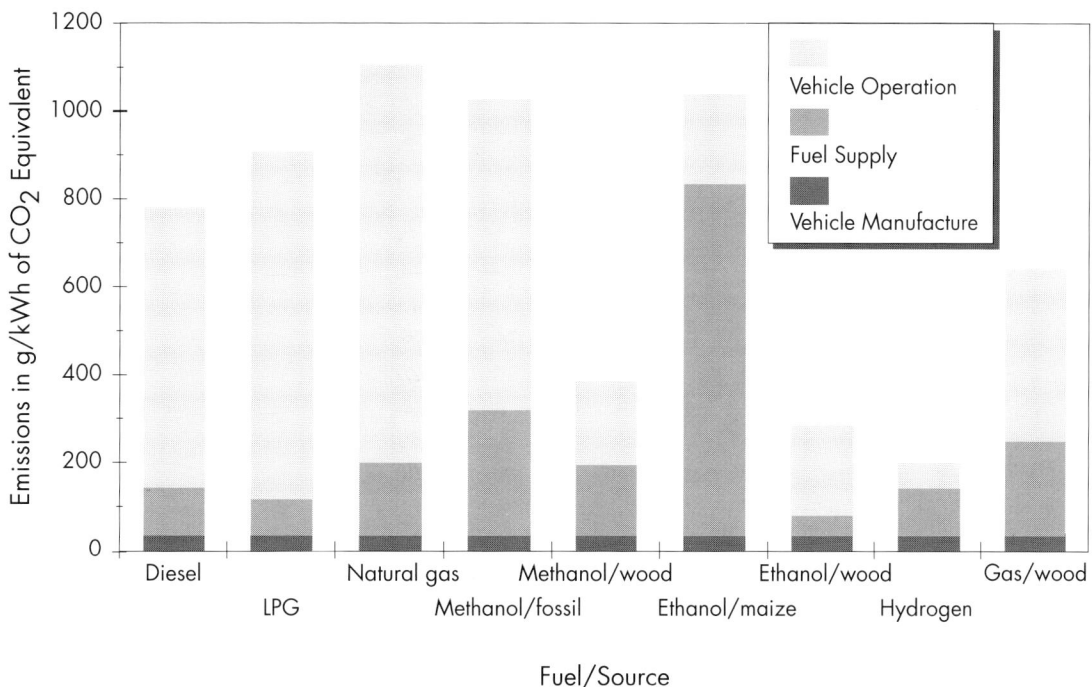

PART III

MARKETS AND POLICIES

CHAPTER 9

ALTERNATIVE FUELS' MARKET POTENTIAL IN NORTH AMERICA

The principal road transport fuel consumed by light-duty vehicles in North America is gasoline. Consumption of gasoline in 1990 amounted to 327 million metric tons. Light trucks make up 20% of gasoline vehicles and are responsible for 30% of gasoline consumption. Most other light-duty vehicles are cars.

Alternative fuels have a very small share in the North American market. The principal alternative road transport fuels are LPG and CNG. Their share (mainly LPG) increased from near zero in 1980 to 0.6% of total road transport energy consumption in 1989.

9.1   THE CURRENT MARKET

9.1.1   Vehicle Stock

Gasoline consumption in North America increased 5% between 1979 and 1990. During the same period, gasoline vehicle traffic rose by 27%, while average fuel economy improved by 17%. The fuel economy of new passenger cars on official tests improved by 30% between 1979 and 1988 (IEA, 1991c). Over the same period, however, there was an expansion in the use of light trucks, which have poorer fuel economy.

Between 1980 and 1988 in the United States, light-duty vehicles with engine capacity between 1 501cc and 3 000cc increased their share of the active fleet from 39.8% to 68.8%. Over the same period, cars of engine size over 3 000cc dropped from 49.7% to 22.4% of the fleet, and those below 1 500cc dropped marginally, from 10.5% to 8.8%. The trend towards smaller vehicles came to a halt in 1989/90 as consumer preference shifted back towards more spacious, powerful and comfortable vehicles.

9.1.2   Gasoline

The share of unleaded gasoline in the North American market varies by region according to the local population of older cars. Leaded gasoline has not been sold in Canada since 1990. While the overall share of unleaded gasoline in the United States is more than 90%, the market share by state varies from 85% in California to nearly 100% in the northeast (d'Zvrilla, 1991).

Since 1988, the volatility of gasoline in the United States has been reduced to cut evaporative emissions. Oil companies have met lower volatility standards by reducing the butane content of gasoline.

The reduction of both lead and butane in gasoline has obliged refiners to seek other means of sustaining gasoline octane levels. They now use octane enhancers such as methanol, ethanol and MTBE. In 1989, this group of oxygenates accounted for roughly 0.8% by volume of the US gasoline pool (PIRA, 1990).

9.1.3   Diesel

Automotive diesel consumption in North America increased 31% between 1979 and 1990. This reflects growth in freight transport. Diesel car sales increased from 1979, peaking in 1981 at around 5% of new car sales. Sales have now returned to low pre-1970 levels, partly because of problems with consumer acceptance and partly because the price advantage of diesel, which during the early 1980s was cheaper than gasoline, has declined and reversed. **Annex 7** discusses the market for diesel.

9.2   OUTLOOK — GASOLINE AND DIESEL CONSUMPTION

In the IEA World Energy Outlook[1] (IEA, 1991b), gasoline consumption in North America continues to level off while that of road diesel rises steadily from 1990 to 2005.

In the Outlook, North American gasoline demand fluctuates between roughly 340 and 345 million metric tons per year over the 15-year period. The fluctuations correspond to changes in gasoline prices and there is no significant upward or downward trend. Slow growth in the vehicle population and traffic is offset by an improvement in fuel economy (see **Figure 9.1**).

The slow growth in the light-duty vehicle fleet reflects a combination of assumptions: consumer expenditure grows more slowly as GDP growth slows; the market approaches saturation; a steady rise in crude oil prices and tightening standards on gasoline and diesel quality contribute to rising fuel prices.

Despite slow growth in the fleet, turnover results in the introduction of more efficient new cars and consequently an improvement in average fuel economy. The increasing number of light trucks in the fleet tends to offset this. Average new car fuel economy is projected by North American governments to improve by 20% between 1990 and 2005 (IEA, 1991b). On this basis the average fuel consumption of the fleet will decrease by about 16% between 1990 and 2005, consistent with the scenario presented in **Chapter 5, Table 5.5**.

---

1.   The Outlook does no take account of the 1990 US Clean Air Act Amendments.

Figure 9.1
World Energy Outlook, North America
(Historical to 1989, Outlook to 2010)

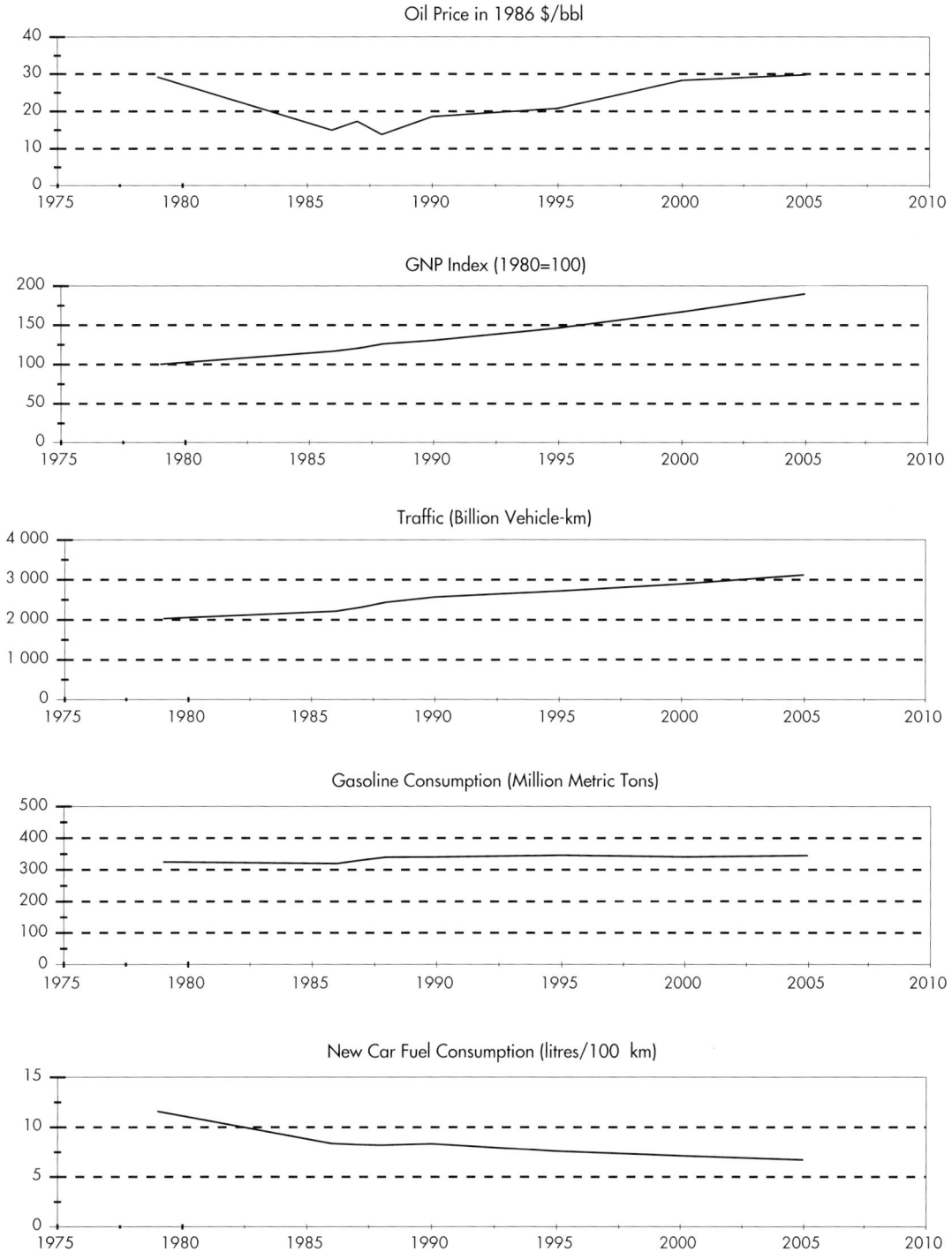

Oil Price in 1986 $/bbl

GNP Index (1980=100)

Traffic (Billion Vehicle-km)

Gasoline Consumption (Million Metric Tons)

New Car Fuel Consumption (litres/100 km)

Source: IEA (1991b)

In the Outlook, North American demand for road transport rises by 53% between 1990 and 2005. The growth reflects increased goods transport as industrial production rises by 2.6-2.8% per year.

9.3 ENVIRONMENTAL LEGISLATION

9.3.1 US Federal Legislation

The most recent amendments to the 1970 US Clean Air Act were signed into law on 15th November 1990. They constitute a significant overhaul of the Act. Pollution control requirements are tightened in cities that are not meeting federal air quality standards, particularly for ozone. The amendments include changes to road transport fuel specifications, and support the use of alternative road transport fuels in those areas with the worst air-quality problems. The amendments are being enforced by the EPA through revisions to vehicle and fuel standards, in negotiation with the automotive industry, the oil industry and the principal environmental groups.

Motor Vehicle Emission Standards. The US vehicle emission standards in force in 1992 are shown in **Table 9.1**. The standards for carbon monoxide, VOCs and NO_x were introduced in 1981. Limits on particulate matter were introduced and tightened during the 1980s. The standards are to be tightened in two phases:

- The Tier I standards will be phased in from 1994 to 1998. They represent a 30% reduction in VOCs and a 60% reduction in NO_x for cars and light trucks. A similar reduction is applied for particulates[1].

- Tier II tailpipe emission standards will come into effect no earlier than model year 2003, and no later than 2006. The precise timing and level of the standards is left to the EPA's discretion.

Table 9.1
US Federal Emission Standards, Light-Duty Vehicles

Model Year	VOC	Carbon monoxide	NO_x	Particulates	Evaporative Emissions (g/test)
		————————(g/mile)————————			
1981	0.41	7.0	2.0		2.0
1982-1986		————no change————		0.6	no change
1987-1994		————no change————		0.2	no change
1994-1996 Limits to be phased in:					
First 50 000 miles	0.25	3.4	0.4	0.08	no change
Second 50 000 miles	0.31	4.2	0.6	0.1	no change
From 2004	0.125	1.7	0.2	0.08	no change

1. Light trucks of the heaviest class face a less stringent standard for particulates; they are not required to meet Tier I until 1999.

The amendments also mandate the EPA to issue regulations concerning cold-start carbon monoxide standards, on-board carbon canisters (to capture gasoline vapours during refuelling), emissions of toxic air pollutants, on-board diagnostics and test procedures.

Fuel Quality. The amendments set out gasoline specifications to be phased in during the 1990s, with the aim of reducing ambient levels of ozone, carbon monoxide and benzene. In 39 metropolitan non-compliance areas[1] for carbon monoxide, gasoline must contain at least 2.7% of oxygen by weight during winter months, since November 1992. Retailers selling gasoline containing less oxygen may buy credits from those selling gasoline with more oxygen. Failure to meet the carbon monoxide air quality standards in a serious non-attainment area by the deadline will result in an increase in the required oxygen content to 3.1% by weight.

Refiners are expected to use gasoline blends containing 15% MTBE. Extensive new MTBE production capacity is being developed around the world to meet demand (CERA, 1992).

Reformulated Gasoline. Beginning in 1995, additional gasoline reformulation will be imposed on the nine metropolitan areas that exceed limits on ambient ozone[2]. These cities account for roughly 25% of US gasoline consumption. Other areas may be included if states so choose and there is enough reformulated gasoline.

Certification standards for reformulated gasoline, which became effective in November 1992[3], are shown in **Table 9.2**. Tradable credits are available for compliance on oxygen, aromatics and benzene content. To avoid refiners transferring gasoline components from one fuel to another, emissions of controlled pollutants from the use of conventional gasoline are not allowed to increase.

Clean-Fuel Vehicle Programme. The amendments established two programmes to promote clean-fuel vehicles: a national fleet programme and the California Pilot Vehicle Programme. The national programme has been strengthened through the 1992 Energy Policy Act, and further measures are under discussion.

The national fleet programme will introduce very clean gasoline or alternative-fuel vehicles into about 25 high-pollution areas. Only fleets of more than ten vehicles fuelled at a central location, and not driven privately at night, are affected. From 1998, a rising percentage of new fleet vehicles in high-pollution areas must meet California standards for low-emission vehicles. If no such vehicles are available in California by then, the programme will be delayed until the vehicles become available or until 2001, whichever comes first. Waivers will apply to emergency, security and rental vehicles.

Clean fuels must be made available at the fleet's central fuelling locations by fuel suppliers. Bankable and tradable credits may be obtained by fleet owners for buying more and cleaner vehicles than required. Fleet programme requirements may be met by converting conventional vehicles to clean fuels.

---

1. A non-compliance area for carbon monoxide is one where the fourth highest recorded concentration in any eight-hour period during the previous three years exceeded the federal standard.

2. Baltimore, Chicago, Hartford, Houston, Los Angeles, Milwaukee, New York, Philadelphia and San Diego.

3. The standards vary according to the time of the year. During the period when ozone levels are highest, VOCs, which are throught to contribute most to ozone formation, are controlled. Reformulated gasoline is to contain the minimum achievable level of VOCs. Benzene emissions must be minimised and heavy metals eliminated all year.

Table 9.2
US Gasoline Specifications:
Current Specifications and Clean Air Act Mandate

Item	Current	Mandate (for non-attainment areas)
Aromatics	30-33%	25% max. by 1995
Benzene	1-3%	1% max. by 1995
Oxygen (summer)	0-1%	2.0% min. by 1992 — by weight 2.7% min. by 1995 — by weight
Oxygen (winter)	0-1%	2.7% min. by 1992 — by weight
Reid vapour pressure	9 psi	Reformulated gasoline
Ozone (high season)	10 psi	Gasohol containing > 10% ethanol
Lead	—	Zero after 1995
VOC	—	Reduce by 15%
NO_x	—	No increase
Total toxic components	—	Reduce by 15%

Note: Since November 1992, gasoline with 2.7% oxygen content has been required in carbon monoxide non-attainment areas (39 cities) during the winter months. Compliance by November 1995 will be imposed on other areas. Year-round compliance will be necessary after 1995 for some cities. State laws may eventually be stricter than federal law.

Source: McKeough (1991).

The 1992 Energy Policy Act makes stronger provisions for the introduction of alternative fuels, focusing on gasoline substitution rather than on vehicle emissions. Targets are set for the conversion of federal and state government fleets and those of companies producing alternative fuels. The aim is to substitute alternative fuels for 10% of gasoline use by 2000, 30% by 2010. Manufacturers are encouraged to produce alternative-fuel vehicles by the provision of Corporate Average Fuel Economy (CAFE) credits related to the amount of gasoline use avoided. Tax credits are available for vehicle conversion and for filling stations (Clean Fuels Report, 1992).

The California Pilot Vehicle Program effectively sanctions existing California programmes. Under this programme, manufacturers must produce at least 150 000 alternative-fuel light-duty vehicles per year for the California market beginning in model year 1996. The number rises to 300 000 in model year 1999. Clean alternative fuels are defined by the Clean Air Act Amendments as methanol, ethanol, other alcohols, reformulated gasoline, reformulated diesel (for trucks), natural gas, LPG and hydrogen or electricity (Clean Fuels Report, 1991). Other states are not empowered to introduce alternative-fuel vehicles. They may provide incentives for the sale of clean-fuel vehicles and clean fuels.

9.3.2 Legislation in California

California's urban air quality problems have long been recognized as more severe than in other regions of the United States. Most of these problems arise in the Los Angeles Basin. The pollution control agency for this region is the South Coast Air Quality Management District

(SCAQMD). The California Air Resources Board (CARB) introduced the Clean Fuels and Vehicles Plan in September 1990. The plan imposes emission standards on new vehicles beginning in 1994 and progressively tightening over the following ten years. VOC emissions must be cut 80% below current levels by 2000. The standards will take effect in Southern California in 1994 and will be imposed state-wide after 1997.

Under the legislation, vehicles will be given one of three new emission classifications, shown in **Table 9.3**. Manufacturers will be allowed to meet a fleet average emission standard by certifying vehicles in combinations of transitional low-emission vehicle, low-emission vehicle, ultra-low-emission vehicle, zero-emission vehicle or conventional vehicle standards. Their sales-weighted VOC emissions must not exceed the prescribed fleet average in a given year. A banking and trading component permits manufacturers to earn marketable low-emission credits (New Fuels Report, 1991a).

Table 9.3
CARB New Light-duty Vehicle Emission Standards
(g/mile)

	TLEV[1]	LEV[2]	ULEV[3]
VOC	0.125	0.075	0.04
Carbon monoxide	3.4	3.4	1.7
NO_x	0.4	0.2	0.2
Benzene	—	0.002	0.002
Formaldehyde	0.015	0.015	0.008

1 Transitional low-emission vehicles.
2 Low-emission vehicles.
3 Ultra-low-emission vehicles.

Note: A zero-emission-vehicle category has been created for vehicles with no exhaust or evaporative emissions; only battery-powered electric vehicles currently qualify.

Source: Clean Fuels Report (1991).

CARB's specifications for reduced summer gasoline volatility, elimination of lead and the use of detergent additives came into force in 1992. Further tightening of gasoline specifications is expected.

Alternative fuels are being promoted by CARB, SCAQMD and the California Pilot Vehicle Program. The fuels are expected to help satisfy low-emission requirements. They include M100, M85, CNG, LPG and reformulated gasoline. At least 90 Southern California filling stations must supply alternative fuels by 1994, rising to 400 by 1997. From 1994, 200 000 new low-emission vehicles per year (about 10% of the state's new car fleet) must be sold in California.

Electric vehicles will be required to satisfy the zero-emission-vehicle requirement. They will have to make up a minimum of 2% of annual new car sales by companies selling more than 30 000 cars per year in California. The share is scheduled to rise gradually to 10% by 2003.

9.4   ALTERNATIVE TRANSPORT FUELS

CNG. There were more than 30 000 CNG vehicles in the United States at the beginning of 1992 and 31 350 CNG vehicles in Canada in 1991 (CERI, 1992; Sypher:Mueller, 1992). All were conventional vehicles retrofitted to operate on CNG. In 1990, Canadian natural gas consumption for road transport amounted to 2.5 PJ (IEA, 1992a). Consumption data for the United States are not available. **Figure 9.2** shows the history of natural gas and LPG use in road transport in North America, as reported to the IEA[1].

Figure 9.2
Use of Natural Gas and LPG in Road Transport, North America, 1960 to 1990
(PJ/year)

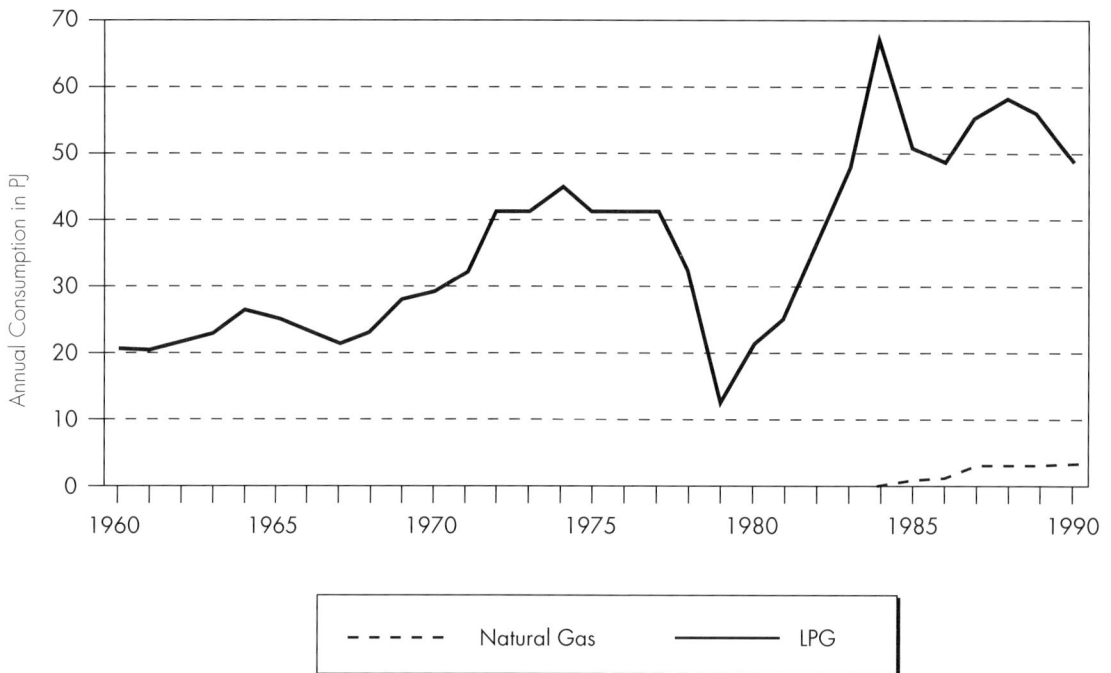

The United States has about 300 CNG filling stations, roughly 6% of them accessible to the public (USDOE, 1988). Most filling stations are owned and operated by gas utilities for their own fleets. Gas utilities were until recently the principal proponents of CNG vehicles, seeking to create a new gas market with less demand seasonality.

A substantial number of US firms market CNG conversion systems for all sizes of gasoline and diesel vehicles, generally with dual-fuel capability (USDOE, 1988). The only dedicated CNG

---

1.   The graph is based on IEA Energy Balances for North America. As gas consumption by transport in the United States is not reported, the gas curve represents an underestimate for the late 1980s.

140

vehicles are heavy-duty vehicles, although experimental versions of light-duty CNG vehicles exist. Gas utilities are helping to finance the development of dedicated CNG cars by major automotive manufacturers.

Until recently, US gas companies concentrated their marketing efforts on fleet vehicles, especially heavy-duty vehicles covering a minimum of 15 000 miles per year. Programmes promoting the conversion of fleet vehicles have been developed in several states, including Arizona, California, Georgia, Ohio, Minnesota, Texas and New York, as well as the District of Columbia (USDOE, 1988).

In Canada, a CNG vehicle programme was initiated after the 1979/80 oil shock. The programme originated in the 1980 National Energy Plan. It was encouraged by the availability of significant reserves of natural gas in western Canada and the general maturity of CNG vehicle technology. In 1983, the Federal Government initiated a system of conversion grants under its Natural Gas Vehicle Program. In addition, two provinces offered sales tax rebates on conversions of new cars, while all provinces offered an overall road tax rebate for CNG. Grants were also made available to subsidise the installation of filling stations. Gas utilities have been closely involved in the promotion of CNG. They have adopted a variety of strategies. Some offered CNG at low prices to encourage vehicle owners to finance conversions. Others charged higher prices for gas and offered rebates on conversions.

Despite government efforts, Canadian market acceptance of CNG vehicles was below expectations. The goal of the 1983 programme was 35 000 vehicle conversions and 125 filling stations. Of the 200 stations approved for subsidies, only 121 were still operating in June 1992. On 31st May 1989, 20 203 CNG vehicles were estimated to be operating in Canada: 42% of them in British Columbia, 32% in Ontario, 21% in Quebec and the remainder in Alberta, Manitoba and Saskatchewan. In marked difference to the United States, 38% of the conversions were on private cars, 25% on vehicles in the service industries, 6-7% each in the manufacturing, taxi and contractor categories and the remainder in other sectors. 41% of conversions were performed on cars, 26% on vans, 25% on pickup trucks and most of the rest on trucks and buses (Energy, Mines and Resources Canada, 1990).

LPG Vehicles. LPG, arising from natural gas and oil production, consists primarily of propane and butane. In the United States, LPG for cars is 95% propane.

North Americans have been converting gasoline vehicles to LPG for more than 60 years. Today it is estimated that there are 370 000 dual-fuel LPG vehicles in the United States (including fork-lifts and off-road vehicles) and roughly 140 000 in Canada. Road transport consumption of LPG in North America was 1.025 million metric tons in 1990, split equally between the United States and Canada (USDOE, 1990b; Energy, Mines and Resources Canada, 1988).

Use of LPG in Canada was promoted by the Federal Propane Vehicle Grant Program, founded under the 1980 National Energy Plan. The programme lasted until the end of 1985, when approximately 140 000 LPG vehicles were on the road, of which 71 000 benefited from grants by the programme. About 90% of the conversions under the programme were done after sale while the rest were done by the original equipment manufacturers. The latter have since withdrawn from the market. Roughly 5 000 filling stations offer LPG refuelling.

The majority of the LPG vehicle conversions occurred in Ontario (46%), followed by Alberta (24%) and British Columbia (19%); these provinces offer a road tax exemption to LPG users. The conversions were predominantly by commercial users. 56% were in trucks and vans, 22% in heavy-duty vehicles and 22% in cars.

Ethanol as a Road Transport Fuel. Ethanol has been used as a transport fuel in North America only in gasoline/ethanol blends known as "gasohol". Its consumption is included in the statistics for gasoline use. The total US consumption of ethanol grew from 1 million cubic metres in 1977 to 4.4 million cubic metres in 1987 and shifted from predominantly industrial uses to predominantly transport fuel uses (industrial demand for ethanol has remained at around 1.1 million cubic metres per year since 1980) (USDOE, 1988). Gasohol accounted for around 8% of US gasoline sales in 1991.

The use of ethanol as motor fuel is promoted in the United States by a variety of agricultural and energy subsidies. Fuels containing at least 10% ethanol from renewable resources are exempt from the federal gasoline excise tax, now $0.06/gallon of gasohol. In addition many states offer fuel tax exemptions ranging from $0.01-08/gallon on gasohol, and some have producer subsidies of $0.20-40/gallon on ethanol. Overall, subsidies amount to $0.25-35/litre of ethanol.

In 1986, total Canadian ethanol demand was about 50 000 cubic metres, of which 15% was used as fuel. Nearly all fuel ethanol in 1986 was produced by the Mohawk Oil Co. at an ethanol fermentation plant (built in 1981) producing 8 700 cubic metres per year. Total Canadian ethanol production was roughly 85 000 cubic metres in 1986. The Federal Government has designated ethanol blended gasoline as an "environmental choice" product. On 1st April 1992, the federal excise tax was eliminated on biomass-derived ethanol for use in blends with gasoline. This amounts to a subsidy of C$0.85/litre of ethanol[1].

Methanol as a Road Transport Fuel. Methanol in North America was used only in chemical feedstocks until the early 1980s. With the phase-out of leaded gasoline in North American markets during the 1980s, demand for methanol grew rapidly. Methanol has been used mainly for the production of MTBE to enhance the octane rating of gasoline in place of lead.

Methanol (in M85 or M100) has been used in about 1 000 vehicles in the United States. Flexible-fuel vehicles accepting these mixtures or pure gasoline are available from North American car manufacturers. In Canada, the abundance of natural gas has encouraged continuing research into methanol for use as a road transport fuel. Methanol production uses some 1.25 million cubic metres, mostly from natural gas. However, only a few dozen vehicles are being tested (USDOE, 1988).

Electric Vehicles. Statistics concerning the number of electric vehicles operating in North America, and their conditions of operation, are limited. A reasonable guess is that there are probably a few hundred on-road electric vehicles, most being used in trials or demonstration programmes.

---

1.   On average in the second quater of 1992, C$1=US$0.827.

9.5   THE FUTURE FOR ALTERNATIVE FUEL AND ELECTRIC VEHICLES

The Clean Air Act Amendments and CARB legislation establish clear guidelines regarding the sale of alternative-fuel vehicles. Canadian legislation is being developed for the promotion of such vehicles and "clean" gasoline and diesel.

California. SCAQMD, CARB and the Clean Air Act Amendments all have designated programmes for the development of alternative-fuel vehicles to improve urban air quality. The programmes focus on reducing ozone and carbon monoxide levels. **Table 9.4** shows estimates by SCAQMD and CARB for the share of alternative fuel vehicles expected in 2010 in the total vehicle-miles travelled in California.

Table 9.4
CARB Projection of Vehicle-Kilometres Travelled for 2010
(% of state total by vehicle class)

Class	Gasoline	Diesel	Electric	Other*
Passenger cars	50%	0%	17%	33%
Light-duty trucks	53%	0%	9%	38%
Medium-duty vehicles	57%	3%	0%	40%
Heavy-duty vehicles	29%	47%	0%	24%
Urban buses	0%	0%	30%	70%

*   Includes methanol, LPG and natural gas.

Source: Clean Fuels Report (1991).

Other States. States can adopt either the federal vehicle standards or California standards. States may also participate in the California Pilot Vehicle Program. The first two states to adopt the California programme were New York and Massachusetts, in late 1990. In May 1991, Connecticut became the third state to adopt the programme (New Fuels Report, 1991a, b and c).

The Clean Air Act Amendments classified US cities according to their attainment of ambient ozone and carbon monoxide levels and gave deadlines for compliance with air quality standards. The categories were: extreme, with full compliance by 2010 (only Los Angeles); severe, with compliance by 2007; severe, with compliance by 2005; serious, with compliance by 1999; moderate, with compliance by 1996; and marginal, with compliance by 1993. Cities in the extreme and severe classifications were in a total of eleven states. Gasoline consumption in these states in 1988 accounted for just over 29% of total US gasoline consumption. States with cities classified as serious accounted for a further 31.5%.

It is not possible at this stage to predict which other states will adopt the California programme. It is also too early to predict the size of the market for alternative-fuel cars that will be stimulated by the programme. However, the legislation is clearly effective in encouraging the demonstration of alternative fuels in the US market.

CHAPTER 10

ALTERNATIVE FUELS' MARKET POTENTIAL IN OECD EUROPE

The principal fuels consumed by road vehicles in OECD Europe are gasoline and diesel. In addition, niche markets exist for natural gas, LPG and electricity. In 1988, nearly 41% of road transport oil consumption in OECD Europe was diesel fuel — more than double the share for North America. This is explained mainly by the large number of diesel cars, vans and minibuses; the virtual absence of heavy-duty gasoline vehicles in Europe; and the high proportion of freight carried by road, in heavy-duty diesel vehicles.

10.1   THE CURRENT MARKET

10.1.1 Vehicle Stock

National average fuel economy figures for new passenger cars in OECD Europe declined between 1979 and 1990, according to official test figures. This decline, however, is significantly less pronounced than that experienced in North America during the same period. In western Germany new car fuel efficiency improved from 9.6 litres/100 km in 1979 to 7.9 litres/100 km in 1990, a 17.7% improvement. Similarly, in the United Kingdom efficiency improved by 17.8%, from 9 litres/100 km to 7.4 litres/100 km. In Italy new car efficiency improved from 8.3 litres/100 km to 6.8 litres/100 km. Other European countries achieved comparable efficiency improvements (IEA, 1991c and **Table 5.2**).

These improvements in fuel efficiency of new cars in Europe were most dramatic between 1979 and 1983. Thereafter they slowed and in the late 1980s reversed. This may be partly due to the fall in real oil prices, which has weakened the economic incentives for car manufacturers to introduce new technology and for consumers to seek fuel efficiency improvements. It is more directly explained, however, by the change in the distribution of car sizes. During the economic boom of the late 1980s, European consumers began to move to bigger, more powerful cars with higher fuel consumption.

The process of introducing catalytic converters into parts of the European market, and the related introduction of unleaded gasoline, began in 1985 — much later than in North America. Since then, countries have been progressively introducing unleaded gasoline and catalytic converters. The share of unleaded gasoline ranges from a few percent in Italy, Spain, Greece and Portugal to over 50% in most of Scandinavia, Switzerland, the Netherlands and Germany (see **Annex 8**). In the European Community, catalytic converters are required on all gasoline car models introduced from July 1992 and all new gasoline cars from 1993.

Until 1992 only a few European countries required new cars to be equipped with catalytic converters. As the lifetime of the average car is about ten years in most countries, the share of cars with catalytic converters in the European fleet is still small.

10.1.2 Gasoline

The rate of introduction of unleaded gasoline depends on availability and price. In the United Kingdom the market share of unleaded gasoline was zero in 1987, 1.2% in 1988, 22% in 1989 and 37% in 1990. The surge in sales after 1988 resulted from a government decision to introduce a tax differential making unleaded gasoline cheaper than leaded, and to oblige filling stations to provide unleaded gasoline. The UK experience is reflected in other European countries. The penetration of unleaded gasoline in Germany is 77%, reflecting a significant price differential and a long-standing policy encouraging the sale of "clean" cars. Spain, in contrast, has a market penetration of 2%, partly because few service stations are equipped to sell unleaded gasoline.

10.1.3 Diesel

In the European Community as a whole, the proportion of diesel cars in the fleet has been estimated at 11% (CEC, 1992). The precise share is not known because some countries do not include vehicle type in their annual licensing data. Sales of diesel cars are accurately known from manufacturers' statistics. The highest proportion of diesel cars is sold in France, where they make up 32% of new cars. Significant market shares in new car sales are also found in Germany (11%), Ireland (14%), Spain (14%) and the United Kingdom (11%).

In some countries the market share of diesel cars is declining. For instance, in Finland diesel cars make up 9% of the fleet but only around 5% of new car sales. A similar reduction is evident in Portugal. Germany shows a slight reduction, with the existing fleet containing 13.5% diesel cars and the new car diesel share at 11%. Although these data indicate that sales of diesel cars have fallen as a proportion of total sales, they do not necessarily mean that the number of diesel cars in national fleets is declining. The lower sales figures may be due to the better durability and lower turnover of diesel cars. An equilibrium has not yet been reached in diesel car turnover.

10.2   OUTLOOK — GASOLINE AND DIESEL CONSUMPTION

In the IEA World Energy Outlook (IEA, 1991b), overall gasoline demand in OECD Europe increases 16.8% from 128 million metric tons in 1990 to around 150 million metric tons in 2005 (see **Figure 10.1**). In the same period, the light-duty vehicle fleet grows at around 2.5% a year. There is a slight decrease in annual vehicle-kilometres and an improvement in the overall efficiency of the fleet.

Figure 10.1
World Energy Outlook, OECD Europe
(Historical to 1989, Outlook to 2010)

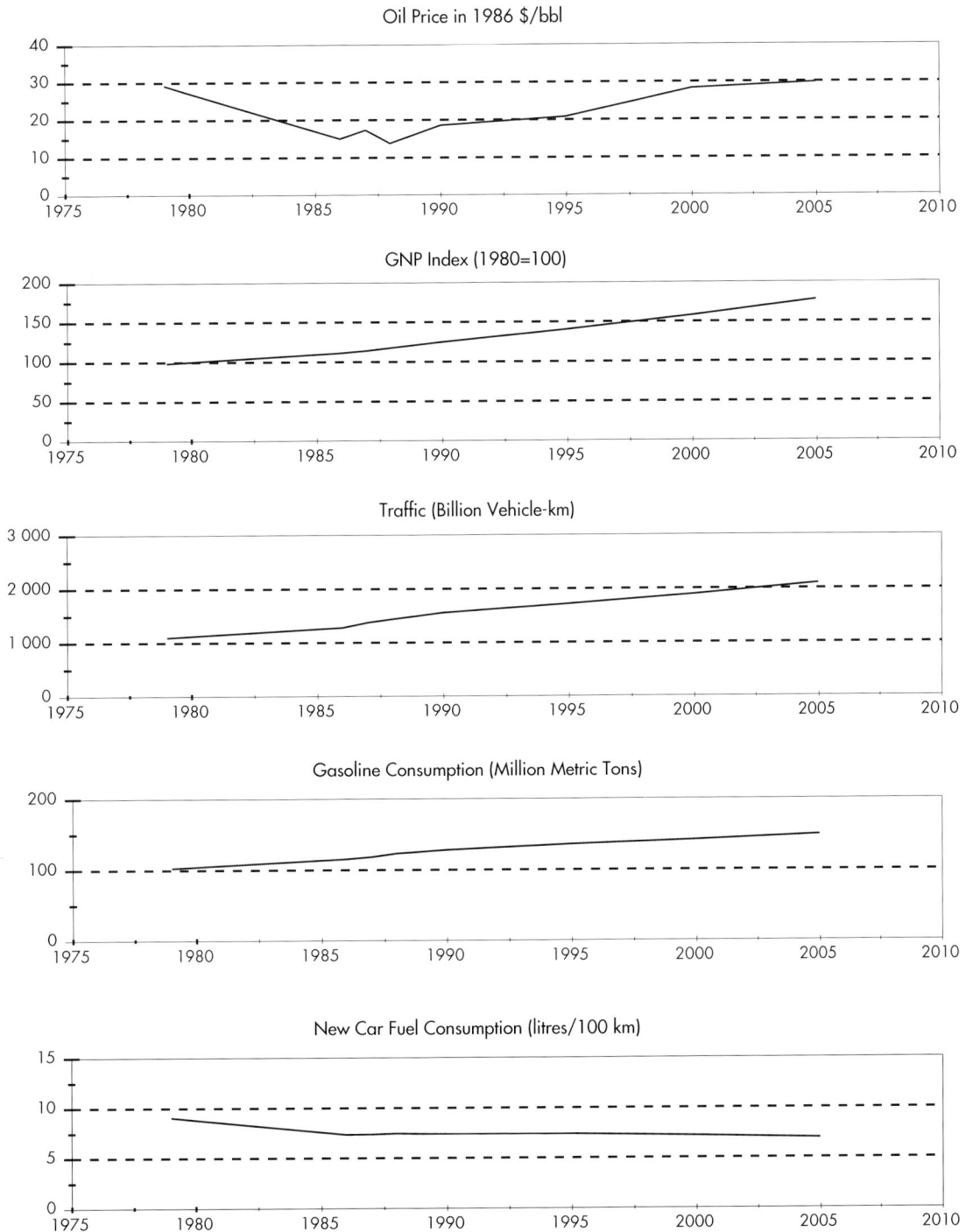

Oil Price in 1986 $/bbl

GNP Index (1980=100)

Traffic (Billion Vehicle-km)

Gasoline Consumption (Million Metric Tons)

New Car Fuel Consumption (litres/100 km)

Source: IEA (1991b).

147

Figures for diesel demand in the Outlook are about 57% higher in 2005 than in 1990. This follows the trend in the late 1980s, when diesel consumption increased significantly (as much as 8.8% in 1988). The rise was largely due to an increase in road freight transport, reflecting growth in manufacturing output. The continuing rise in the proportion of diesel cars also contributed.

The Outlook assumes that no significant modal shift develops between 1990 and 2005. It assumes that the single EC market increases levels of transport and improves efficiency, due to industry liberalisation. It does not take into account growth in the eastern European economies and demand on western European transport systems. The latter factor could significantly accelerate expansion of fuel demand for road freight transport in OECD Europe.

Because few national data on the breakdown of fuel use are available, the Outlook does not differentiate between consumption of diesel fuel by passenger cars and by heavy-duty vehicles. There is significant potential for an increase in the demand for light-duty diesel vehicles. The supporting indices for the Outlook show a growing divergence between the prices of gasoline and diesel fuel.

10.3   ENVIRONMENTAL LEGISLATION

10.3.1 Legislation in the European Community

Cars in EC countries and several other European countries, including some in eastern Europe, are built to standards defined by the UN Economic Commission for Europe (UN-ECE). The standards have been incorporated into successive directives issued by the EC Council of Ministers as the most stringent standards that Member states may apply.

The original UN-ECE standard for motor vehicle emissions, known as Regulation 15 (R15), was included in the 1970 EC Directive 70/220/EEC. R15 is based on a standard dynamometer test cycle, known as the R15 Type I Test, which is intended to simulate urban driving conditions. The standard limited carbon monoxide and hydrocarbon emissions to levels that depended on the weight of the vehicle. For each pollutant a "type approval" level was defined for new vehicle models, as was a "conformity of production" level against which a sample of vehicles from a given production line would have to be checked.

The limit values were reduced in successive amendments to R15, the final version being R15.04. NO_x emission standards were included in the later amendments. The amended standards were translated into EC directives. Member states ratified successive amendments in nationally regulated standards, although the date of ratification (and hence the amendment in force at any time) differed between countries.

Directive 91/441/EEC, known as the Consolidated Directive, simplifies emission standards for all new vehicles below 2.5 metric tons, of whatever fuel and engine type, bought in the Community. It is designed to force the use of catalytic converters or equivalent emission controls on gasoline cars, and at the same time to permit the continued use of diesel cars. To satisfy both these requirements, NO_x and VOC emissions are limited in total, rather than separately. The directive also includes requirements relating to crankcase and evaporative emissions and the durability of emission control systems.

The Consolidated Directive defines a new test cycle, combining the R15 Type I Test and an extra-urban driving cycle. This provides a better indication of NO_x emissions, which are correlated with speed. The limit values for the various pollutants are set out in **Table 10.1**. Limit values for particulate emissions from diesel engines were introduced by Directive 88/436/EEC and updated in the Consolidated Directive and are shown in the table.

Table 10.1
Limits in the Consolidated Directive
(g/km)

Emission Type	Limits Type Approval	Conformity of Production
Carbon monoxide	2.72	3.16
VOC + NO_x	0.97	1.13
Particulates	0.14	0.18

Note: Evaporative emissions are limited to 2g/test. Crankcase emissions are required to be zero.

Source: EC Directive 91/441/EEC.

Previous directives, which left the implementation of limit values to the decision of individual governments, were related chiefly to the aim of eliminating trade barriers (CONCAWE, 1991). The Consolidated Directive is motivated far more by environmental concerns. It is mandatory for EC Member states, which were required to bring into force the laws and regulations necessary to comply with the directive by 1st January 1992.

In addition, the directive required the Commission of the European Communities to make recommendations on measures designed to:

- limit CO_2 emissions;

- adapt the emission standards to vehicles not covered by the directive (including all commercial vehicles);

- lay down regular inspections and procedures for replacing, repairing or maintaining equipment fitted in order to meet the limit values;

- implement a research and development programme to encourage the marketing of "clean" vehicles and fuels.

The sulphur content of diesel fuel in EC countries is limited by Directive 87/219/EEC to 0.3% by weight. In Germany and most of the Stockholm Group (Austria, Finland, Norway, Sweden and Switzerland) the limit is 0.2%. In Denmark, which relies on economic incentives, urban buses use a "city grade" diesel (sulphur content less than 0.05% by weight).

A further EC directive will reduce the sulphur limit to 0.2% starting on 1st October 1994 and to 0.05% by 1st October 1996. Diesel fuel at the 0.05% sulphur level will have to make up at least 25% of sales in EC countries starting on 1st October 1995 (CEC, 1991).

The European Commission has produced proposals for emission standards to come into effect in 1996. These are shown in **Table 10.2**.

Table 10.2
European Commission Proposal for 1996

Emission Type	Gasoline Cars	Diesel Cars	
		Indirect Injection	Direct Injection Until 30.9.99
Carbon monoxide	2.2	1.0	1.0
VOC + NO_x	0.5	0.7	0.9
Particulates		0.08	0.1

Source: Europe Environment (1993).

10.3.2 Legislation in Non-EC Countries

OECD countries in Europe outside the EC have varying legislation on emission standards. Most have adopted some hybrid of the UN-ECE standards and US standards. In some cases this involves applying US emission limits (on a g/km basis) in the R15 Type I Test and the US FTP highway driving cycle.

10.4 ALTERNATIVE TRANSPORT FUELS

Total road fuel consumption in OECD Europe in 1990 was 226.99 Mtoe (IEA, 1992a). Gasoline and diesel accounted for 224.26 Mtoe, LPG for 2.52 Mtoe (about 1%) and natural gas for 0.21 Mtoe (about 0.1%). In 1980, total road fuel consumption was 166.4 Mtoe, including 1.47 Mtoe of LPG and 0.26 Mtoe of natural gas. Consumption of other alternative fuels and electricity was minimal.

CNG. Italy has used CNG in vehicles since the 1930s. During World War II, CNG was tested as an alternative fuel in France, Germany and Britain. Italy and France continued their programmes after the war. The Netherlands also has a substantial natural gas vehicle programme. Except in Italy, however, very little CNG is used in OECD Europe road transport today (USDOE, 1988). Italy is the predominant supplier of all types of CNG vehicles, conversion kits and refuelling equipment.

CNG consumption in Italy declined during the 1960s, then increased following the oil price shock of 1973. The 1986 National Energy Plan called for more natural gas use in the transport sector. In 1990, road transport consumption of natural gas in Italy was almost 0.21 Mtoe, accounting for virtually the entire consumption of natural gas by transport in OECD Europe. Even so, it amounts to only about 0.7% of Italian road sector energy consumption, the rest being oil products. **Figure 10.2** shows the history of natural gas and LPG use in road transport in OECD Europe.

Figure 10.2
Use of Natural Gas and LPG in Road Transport, OECD Europe, 1960 to 1990
(PJ/year)

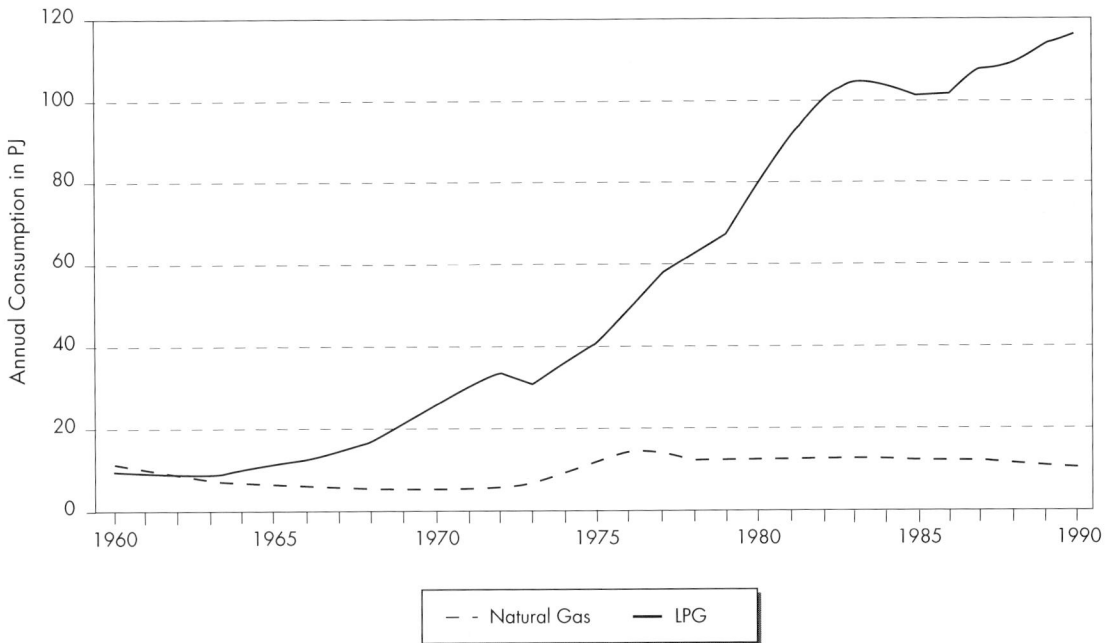

Source: IEA Energy Balances.

There are about 235 000 CNG-fuelled vehicles in Italy, mostly dual-fuel private cars and some trucks. This represents about 1% of the total number of cars in Italy. In addition, several diesel city buses owned by local governments have been converted recently. These vehicles are concentrated in northern and central Italy, where there are about 240 fuelling stations. CNG stations account for 0.7% of Italy's total refuelling network for all transport fuels. Their limited number and geographical spread has been one of the main constraints on the growth of CNG vehicles in Italy. CNG is favourably priced compared to other fuels in Italy, at less than one-third the price of gasoline and half the price of diesel or LPG, in terms of energy content. The price differential has clearly encouraged the use of CNG.

Italy has strict regulations governing conversion to CNG. Cylinders must be tested every five years. To encourage compliance, damaged cylinders are replaced free of charge. The cost of the tests is met from an annual fee paid by cylinder owners, which also provides liability insurance against cylinder filling accidents (USDOE, 1988).

CNG is being demonstrated in several other European countries, including the Netherlands and Sweden. One market niche expected in these countries is for urban buses. In the Netherlands about 20 CNG buses have been demonstrated. The interest in alternative bus fuel arises from health concerns related to NO_x and particulate emissions.

The Netherlands had about 200 CNG cars in 1991, and the target of 1 000 for 1992 is thought to have been exceeded (Okken, 1992). Most are used in the Amsterdam area, where there are three public filling stations. Most CNG car owners rent a home compressor and use their domestic natural gas supply.

LPG. The consumption of LPG as an automotive fuel in OECD Europe exceeds that of CNG. LPG is used as a transport fuel in most of the countries in the region, except for Finland, Portugal, Switzerland and Turkey. In Portugal the use of LPG for road transport is forbidden. In Finland and Switzerland, high taxes have discouraged its use as transport fuel.

Interest in LPG as a transport fuel is motivated partly by the lack of an alternative high-value market in many countries. Recently, however, it has received attention in Europe as having less environmental impact than gasoline, and has been promoted as such in the Netherlands and Italy.

LPG is the predominant alternative fuel in the Netherlands and Italy. Supplies there are abundant, partly because of the size of the two countries' refinery industries. In 1990, nearly 908 000 metric tons of LPG (1.03 Mtoe) was consumed in the transport sector in the Netherlands, and 1.34 million metric tons (1.52 Mtoe) in Italy. LPG's share of road transport energy use was about 12.5% in the Netherlands and 5% in Italy (IEA, 1992b). The two countries are responsible for 92% of road sector LPG consumption in OECD Europe. Road transport use of LPG in OECD Europe grew by 4.4% from 1989 to 1990, while total road transport energy use grew by 3.4%. The only country with significant growth in LPG use was Italy, with a 7.5% increase. In most other countries, including the Netherlands, LPG use declined.

Biofuels. Biofuels for road transport are not produced to a significant extent in Europe. The CEC's Forward Studies Unit (Cellule de Prospective) recently analysed the economics of various options for biofuel production (Wright, 1991). The study looks at a range of fuels, including transesterified rapeseed oil, ethanol from wheat and sugar beet, and lignocellulose used in direct combustion. According to the report production of any of these fuels is likely to have a positive impact on employment, the Community budget and, potentially, the environment.

The analysis led to a Commission proposal to encourage the use of biofuels by tax exemptions. The proposal suggests that the fuel duty on biomass products should be limited to 10% of that on petroleum products in the same application. The effective subsidy for biofuels would be comparable with the $0.25-35/litre incentives for ethanol use in many states in the United States. France already has a tax exemption for biomass-derived fuels. The proposal would have its greatest effect on transport fuels, where duties are highest.

Conversion of wheat and sugar beet to ethanol are under consideration, and there are research, development and demonstration programmes on the relevant technologies, funded both by European governments and by the Commission.

OECD Europe has an annual wheat surplus of about 20 million metric tons, mostly from EC countries. It could be converted to about 6 million metric tons of ethanol, or 4 Mtoe. Ethanol could be used in European gasoline in a 5% blend. Europe's only plant producing ethanol from wheat is in Sweden.

Methanol. Much research has been undertaken in Germany on the blending of methanol with gasoline. The automotive industry, fuel suppliers and the German Ministry of Economics reached an informal agreement to introduce a 3% methanol blend, which now holds about 80% of the gasoline market. Its market share, however, has been recently threatened by the availability of cheaper MTBE (USDOE, 1988).

Rapeseed Dimethyl Ester. There is a strong interest in OECD Europe in the use of vegetable oils as feedstock for transport fuels. At present the fuel receiving the most interest is RME, obtained by a transesterification process that involves a reaction of rapeseed oil with methanol in the presence of sodium hydroxide. The process also results in two principal byproducts, glycerol and a protein-rich cake that can be used as animal fodder. The market for the protein is an important determinant of the economics of the process. Glycerol can be also sold, but the market is too small to absorb the amount that would be produced from widespread RME production.

RME is a direct substitute for diesel fuel, requiring no engine modifications and having essentially the same characteristics. There have been claims that diesel engine emissions are substantially reduced with this fuel, which is sulphur-free. However, there does not appear to be a significant difference in emissions between RME and low-sulphur diesel. Studies of the energy balance for RME indicate that the fossil fuels used in agriculture and processing are equivalent to at least half the diesel fuel displaced. Fossil fuels in processing could be replaced by rape straw, although this has not been done in existing demonstration plants.

Electric Vehicles. Research into electric cars surged in Europe during the 1970s and several unsuccessful prototypes were produced. The zero-emissions legislation in California has motivated more intensive work on electric cars by European manufacturers, resulting in several demonstration or "concept" cars and some production models. Many of the cars are hybrids.

Unlike research into alternative fuels, much of the original work on electric cars has been concentrated in industries that are not traditionally involved in car design. Development of vehicles has been undertaken by consortia of battery, electric motor, electronics and vehicle manufacturers. In some cases vehicle manufacturers have acquired electric drive-train component manufacturers.

The German Gesellschaft für Elektrische Strassenverkehr (GES) was founded in 1970 to do research and development work on electric vehicles. Participants included Rheinisch-Westfälisches Elektrizitätswerk (RWE, the electric power utility) and battery and car manufacturers. More than 300 electric vehicles were manufactured and tested as a result of this work. In 1987 the GES was dissolved and RWE took over the electric vehicle R&D programme (Institute of Applied Energy, 1991).

In France, a number of the smaller car manufacturers have been developing electric vehicles since the 1970s, with the major manufacturers, Peugeot and Renault, developing and demonstrating prototypes. Electricité de France and the PSA Group (Peugeot & Citröen) have an agreement to manufacture and deliver 250 electric vehicles between 1990 and 1994. In an attempt to reduce production costs, the cars will be built on the same production line as internal combustion vehicles. Eighty units have been produced and are on the road. The French Government provides some support for electric vehicle development.

In Italy, Fiat recently announced that it would begin commercial production of the Panda Elettra. The Progetti Gestione Ecologiche is setting up an electric vehicle rental system. Both projects are partially subsidised by the Government.

CHAPTER 11

MARKETS AND POLICIES

This chapter explores ways policy makers can respond to the environmental problems caused by car use. The principal questions addressed relate to whether alternative fuel and more efficient technology might achieve a larger market share and, if so, how. These questions should not be answered solely in relation to the issue of climate change. Governments have many reasons for considering the promotion of new technology and fuel. Existing programmes often aim to improve energy security and reduce urban air pollution. Greenhouse gas emission abatement is likely to be an additional benefit of technology and fuel changes rather than the sole reason for their adoption. Conversely, where alternative fuels and electric vehicles result in higher greenhouse gas emissions, this must be balanced against benefits such as reduced urban air pollution. It is difficult for policy makers to find a balance, especially without reliable information on the external costs associated with greenhouse gas emissions. Some analysts believe such costs cannot be calculated.

An ideal transport policy would eliminate the problems due to cars while maintaining all the benefits associated with car ownership and use. Although this can be a long-term goal, it is not yet possible. In the next few decades, the best that might be attained is the identification and internalisation of costs associated with car use, resulting in a more efficient trade-off between the benefits of car use and the burdens on society.

11.1   CONDITIONS FOR TECHNICAL CHANGE

Even technology that appears financially attractive sometimes fails to achieve a significant market share without government intervention. Several players are involved in the introduction of an alternative fuel: fuel suppliers and distributors, vehicle manufacturers, consumers and the government. This section examines their roles — in particular those of manufacturers and consumers — and the barriers that need to be overcome to obtain their participation in alternative fuel programmes.

11.1.1 Vehicle Manufacturers

Gasoline remains cheap and plentiful and gasoline engines are the cheapest way of providing power for most personal private cars, even where fuel and vehicle taxes are high. Major car manufacturers rarely decide of their own accord to market alternative fuel vehicles, although they do produce prototypes. The main exception has been the introduction of diesel vehicles in Europe in recent years (see **Annex 7**). Manufacturers have undertaken major research programmes into alternative fuels and electric vehicles, often with government support.

The reluctance to commercialise new car technology can be explained by a number of factors. Most alternative car types are not yet ready for the market. Even where technologies are sufficiently mature for commercial mass production, it may not be clear that the market is ready for them. There is also some reluctance on the part of manufacturers, partly because they believe that car purchasers are very conservative and partly because they are hesitant to encourage substitution away from well-established products in which they have considerable investment.

While a few small companies do market alternative-technology cars, especially electric cars, it is very hard for them to compete with the conventional manufacturers. The level of investment required for low-cost vehicle production and testing, as well as the established dealership networks and brand images of the major companies, poses significant barriers for new entrants to the market.

Technical Risk. In many instances further development is needed to produce a marketable product and manufacturers may be unwilling to take the risk involved in funding this work. In other cases there is simply no prospect of the alternative vehicle being financially competitive with conventional cars.

Governments have generally responded to manufacturers' unwillingness to invest in research by co-financing projects. One difficulty for policy makers lies in deciding which programmes to fund and how much money to offer. Governments have developed a variety of means of auditing programmes, but industrialists tend to be better informed than civil servants on the value of different lines of research, and it is hard to obtain neutral comparisons.

It is difficult for governments to intervene directly in technology development beyond fairly basic science, except where companies are state-owned. The closer a technology comes to being marketable, the less willing companies are to share research and development results.

The government strategy in California involves establishing emission standards ten years ahead. Requiring zero-emission vehicle sales effectively instructs manufacturers to develop a specific technology. Manufacturers failing to work on electric vehicles may be incurring more risk than those who invest in this technology.

Market Risk. Cars form a remarkably homogeneous group of products. Production costs are minimised by the use of a given engine block or chassis for several models. Car component factories have very high production rates. A typical car factory produces some 100 000 cars a year. Economies of scale and automation are very important in keeping costs down.

Some costs associated with producing a new car model are unavoidable. While a production line for 100 000 cars can cost $1 billion, the design of a new model can cost about $100 million. The design phase has to be carried out regardless of the number of cars to be produced. Since there are important economies of scale in the production line, specialist cars produced in small runs can cost over $100 000 each. Vehicles at this price clearly would not begin to penetrate the mass market.

Optimised alternative-fuel vehicles will have to start with completely new designs — for example, with altered structures to deal with the weight of fuel storage systems. The cars may have to be produced in short runs until a viable mass market is demonstrated. This will result in high production costs. Prospective buyers are unlikely to want these vehicles at high prices, especially when they cannot be sure of the vehicles' performance, reliability and resale prospects, or of the relative prices of fuels.

Alternative-technology vehicles can be introduced without imposing all these costs on the first buyers:

- by retrofitting equipment and making minor modifications to existing vehicles and engines;
- by large manufacturers who can spread the development and investment cost over the rest of their product range;
- by manufacturers who are prepared to take the risk of installing full-scale mass production facilities for the alternative technology.

All these approaches have been taken. The most common is the use of retrofits on finished or nearly finished standard vehicles. This is cheaper than changing production lines. Conversion kits can be mass produced to fit on a range of existing engine types. **Chapters 9** and **10** mention a number of retrofit programmes in OECD countries.

The second option is likely to be adopted by manufacturers wishing to sell vehicles in California after 1998. The development and production cost of electric vehicles will almost certainly exceed the prices consumers are willing to pay for them. As manufacturers are required by law to sell such vehicles, they may have to offer them below cost and make up the difference through the rest of their sales.

The third option, if it succeeds, can result in the alternative technology being produced at competitive costs. However, there is a substantial risk to the manufacturer. Early experience in the Brazilian ethanol programme demonstrated the difficulty of persuading manufacturers to provide dedicated ethanol vehicles. Large-scale production facilities were set up only after the government offered long-term fuel price guarantees, which effectively assured the manufacturers of a market.

In some instances, notably in New Zealand and Brazil, alternative-fuel programmes have depended on government subsidies and tax differentials. These programmes were expensive, and lost impetus when the governments reduced support.

Manufacturers will not lay out the investment for new production lines if they are not sure of a market for the product — they express particular concern regarding the lack of supply infrastructure for alternative fuels. It is understandable that no car manufacturer is prepared to take the risk of being the only one to develop an alternative-fuel vehicle.

Market Strategy. Car manufacturers are prepared to invest heavily to be the first to market improvements to gasoline vehicles. They do not seem so interested in being the first to develop and market alternative technology. Any switch to alternative fuels and, especially, to electric vehicles could reduce the market share of conventional car manufacturers. They would have to rely increasingly on the products and expertise of other industries.

Regulations can be used to encourage industry to adopt improved technology without governments "picking winners". The California law on low-emission cars has gone some way towards doing this; it does not necessarily require the use of alternative fuels to reduce emission levels. Car manufacturers have achieved the low-emission requirements using improved conventional technology in gasoline vehicles. The zero-emission requirement, on the other hand, gives manufacturers no option but to produce electric vehicles.

Lack of Standards. Until recently few governments have set separate emission and safety standards for alternative-fuel vehicles. Any manufacturer finalising a car design before standards are set would be taking the risk that the car would not meet future standards.

Emission standards developed for gasoline cars may be inappropriate for alternative fuels. In the European Community standards are flexible enough to be met by both gasoline and diesel vehicles. US fuel and car producers have been concerned that ethanol/gasoline mixtures and CNG will not meet future emission standards. The addition of ethanol raises the vapour pressure of gasoline, taking it beyond EPA standards for reformulated gasoline. CNG vehicles have high total hydrocarbon emissions, of which about 90% is methane. Recognising these impediments, the EPA has set standards allowing for higher vapour pressure in fuels with ethanol, and standards for CNG vehicles cover only their non-methane hydrocarbon emissions. These "relaxed" standards give suppliers a strong signal that the benefit of introducing the alternative fuels is considered to outweigh some minor environmental disadvantages.

Alternative fuels raise different safety issues to those encountered with gasoline. Gaseous fuels carry the obvious risk of cylinder failure. Standards and test procedures are required for on-vehicle storage and for distribution facilities. Some alternative fuels, such as methanol, are more toxic or more easily assimilated into the body than conventional fuels. The traditional uses of ethanol and MTBE have been for their inebriating or anaesthetic effects. Thus they may require special controls.

11.1.2 Fuel Suppliers

Most of the fuel for existing cars is supplied by a small number of oil companies. In general, these companies have complied with government decisions to change fuel standards, while pointing out the cost of doing so. In many instances government regulations or incentives have been effective — for example, in introducing unleaded gasoline and cleaner diesel fuels. The oil companies have treated these situations as opportunities to gain market share.

Because of the potential competition posed by non-petroleum fuels, most oil companies have done some research into alternative fuel technologies ranging from oilseed production to coal gasification and gasoline synthesis. Some of these processes have been deployed, but it is unlikely that oil companies would promote alternative fuels before market conditions, including the regulatory and fiscal frameworks, made it attractive to do so.

Gas and electricity suppliers have been quick to recognise the transport sector as a potential new market. They have become the strongest proponents of CNG and electric vehicles. Natural gas producers have been the organisations best placed to guarantee the availability of CNG at an attractive price. They have also taken responsibility for putting CNG vehicles on the market, to the extent of co-funding gasoline car conversions and, in one case, a production line for dedicated CNG cars. Similarly, electric utilities and their trade associations are the main promoters of electric vehicles, and the main proponents of biofuels are the farm lobbies.

These groups, however, are not in a position to fund vehicle production or conversion. The high cost of electric vehicles is only partially offset at present by lower energy costs. Biofuel production costs are higher than those of petroleum-derived fuels, so again there is no possibility of offsetting vehicle production costs. Biofuels have not entered the market in any quantity without government intervention. Electric vehicles are likely to remain in a similar position for some time.

11.1.3 Consumers

Cars are principally a consumer good. Most are sold to consumers for private, personal use. The main hurdle for any new car technology is consumer acceptance.

Consumers respond to many characteristics of cars. The decision to buy depends on the match between the vehicle and the consumer. **Table 11.1** lists some factors affecting prospective vehicle buyers. Fuel economy is generally a relatively low priority. Convenience — the range between refuellings, ease of finding refuelling stations, reliability, ease of maintenance — is a very high priority. Reasons that consumers may not buy a new type of car include:

- lack of information on running costs, resale value, performance, environmental impact and vehicle availability;

- a perception of risk, especially with gaseous fuels, due to unfamiliarity with the technology;

- real risks, such as inadequate proof of safety or an unpredictable resale market;

- a mismatch between the car's performance and costs and the consumer's needs.

Table 11.1
Factors Affecting the Decision to Buy a Car

Car Attributes	Buyer Attributes
Purchase price	Company or private purchase
Expected resale value	Income
Performance and handling	Number and type of cars already owned
Interior volume	Gender
Expected running costs	Marital status
Fuel availability	Type and location of employment
Expected reliability	Location of home
Vehicle styling, colour, etc.	Leisure activities
Perceived safety	Social status
Image	Size of garage
Brand name	Number and age of children

Lack of information is relatively easy to remedy. Misperceptions can be harder to correct. Real failures to meet consumer needs may be impossible to correct. Some barriers can and should be dealt with by the producers of the technology.

Information. If a car is marketable, suppliers are likely to advertise it. Public research and information programmes can help by providing an authoritative and neutral view of the options open to consumers. They can also help familiarise consumers with new technology and reduce undue perceptions of risk. The most important role for government is in the formulation of standards for product information.

Safety Standards. The establishment of safety standards and test procedures can be important in improving confidence in new technology, but risks intrinsic to the technology are harder to deal with. The differences between the safety hazards posed by CNG or electric vehicles and those associated with gasoline and diesel will be fully understood only after several years of widespread use.

Market Uncertainties. A more important problem for the introduction of new car technology has been market risk. Consumers are unlikely to buy a new car if they are unsure of the resale market or if the future cost and availability of the fuel is uncertain. Governments wishing to promote alternative fuels have found ways of stabilising the market, and this turns out to have been a key to the apparent success of many alternative fuel programmes. Many governments have provided guarantees that the alternative fuel price will remain well below that of gasoline. The durability of such guarantees has been particularly important in maintaining consumer confidence. Similarly, fuel availability has had to be guaranteed, generally by requiring some service stations to provide alternative fuels.

The approach being adopted in the United States is of interest partly because it does not involve government intervention on fuel price. Natural gas is much cheaper than gasoline and there appears to be little risk of this position changing in the medium term. The recent surge in CNG use in the United States is attributable mainly to the progressive tightening of vehicle emission standards expected over the next ten or more years. Consumers and fleet operators buy CNG vehicles because they believe that the emission standards will assure a future market.

Desirability of Technology. It is hard to promote new technology that does not meet consumer requirements. In the car market, one of the most important recent examples is the catalytic converter. These devices satisfy a social need for clean air but compromise some driver requirements. They reduce engine power and efficiency slightly. They also depend on the use of unleaded gasoline, which costs more to produce than leaded gasoline. It is widely recognised that where catalytic converters have been introduced without higher taxes on leaded gasoline than on unleaded, misfuelling is common (see **Annex 8**). Most countries that require cars to be fitted with catalytic converters also introduce a tax differential between leaded and unleaded gasoline.

Implications for Alternative Technology. Chapter 7 compares the projected costs of operating gasoline, diesel and CNG vehicles in 2000. It shows that drivers in the United States could save up to 1¢/km by switching from gasoline to CNG. This saving over 10 years' average driving is equivalent to a car cost reduction of $1300 at a 5% discount rate. The financial saving may not be enough to outweigh the disadvantages of CNG cars. These will include the short range on a full tank, difficulty in finding fuel, long refuelling times and difficulty in finding a buyer when the time comes to sell the car.

Alternative fuel technology that compromises vehicle range or performance is likely to have its market restricted to buyers who do not value those attributes highly. This is one reason many demonstration programmes have relied on sales to commercial fleet operators, whose primary objective is minimising cost. Within the personal car market, drivers with higher annual travel tend to be those who most value comfort, performance, convenience and other status-related attributes. Although these drivers may have the greatest financial incentive to use alternative fuel and energy-efficient technology, they are also relatively unlikely to respond to the incentive.

The image factor can be made to work for alternative technology. A market study in California indicates that, all other attributes being equal, consumers are attracted by the image of alternative fuel and electric vehicles (Golob et al., 1991). This might be expected of a population that has been exposed to considerable publicity regarding the benefits of alternative vehicles. And yet the study indicates that other vehicle attributes, including range, fuel availability and performance, are likely to outweigh the image effect.

11.2   INTERVENTION IN THE TRANSPORT MARKET

Despite the difficulties, governments and alternative fuel producers may have a rationale for promoting alternative fuel and electric vehicles where they can improve energy security and reduce pollution or other problems caused by cars. **Table 11.2** presents an inventory of policy measures that can affect greenhouse gas emissions. They are divided into fiscal measures (taxation and government spending), regulatory measures and information.

Several types of decisions by car buyers and drivers will affect emissions. They include choice of car, frequency and quality of maintenance, choice of fuel, distance driven, route, load carried, timing of trips and driving style. Some of these decisions affect the emission rate (greenhouse emissions per unit of vehicle travel), some affect the amount of traffic, some affect both. The rest of this chapter examines the case for government intervention in the transport sector and its possible effects on the alternative fuel and electric vehicle markets.

The transport market is heavily taxed and regulated in most OECD countries. Rationales for the taxation vary. Cars were viewed as a luxury in Europe until the 1960s. For most countries they are an imported luxury and require imported fuel. Governments partly justify car and fuel taxes by the desire to improve trade balances. Car industries in many producer countries have been, or still are, state owned, and protected by import quotas and tariffs. In some countries car and fuel taxes are justified as revenue to cover government expenditure on roads, police and other related services. The taxes are an effective way to raise revenue because the price elasticities of car and fuel use are quite small.

The imposition of higher taxes on gasoline than on diesel fuel has been important in encouraging a diesel car market. In some countries the main policy objective is to limit the tax burden on freight transport. The promotion of diesel cars for personal use was a secondary, unintended effect. More recently, differential taxation has been important in promoting technology with less environmental impact. The example of unleaded gasoline has already been mentioned. The relatively high level of fuel taxation in Europe offers considerable opportunities to promote other fuels through tax exemptions. There is a wide variety of tax exemptions for alternative fuels in place in both North America and Europe. These measures have encouraged some adoption of alternative fuels, as discussed in **Chapters 9** and **10**.

Table 11.2

Policy Measures to Influence Passenger Transport Supply and Demand, and their Areas of Impact

Area of Impact	Economic Measures	Regulatory Measures	Information
1. Fleet growth	- Annual fees for "right of access" to a vehicle - Vehicle purchase taxes - Fuel taxes - Annual ownership taxes - Parking fees - Road use fees or kilometrage charges	- Annual sales quotas - Individual ownership limits (e.g. total number, no car without a parking space) - Closing some urban districts to cars - Limits on elective or professional travel - Limits on numbers of parking spaces - Increasing congestion by not improving infrastructure - Better infrastructure for bicycles, pedestrians - Expanded/improved public transport and lanes restricted to certain users	- Public information on mode cost comparison
2. Fleet mix	- Taxes and rebates on vehicle purchase determined by size, power or fuel consumption - Fuel taxes - Annual registration taxes determined by engine size - Manufacturer fiscal incentives for fuel efficiency, or fees for non-compliance with standards	- Minimum fuel efficiency standards (e.g. CAFE) - Maximum emissions standards - Maximum power/weight ratios - Civil penalties for manufacturer non-compliance with standards or requirements - Annual sales quotas (by vehicle size)	- Public advertising promoting clean, fuel-efficient small cars
3. Unit emissions	- Fiscal incentives for successful vehicle inspection, or non-compliance fees - Fiscal incentives to retire old vehicles, or non-compliance fees - Fiscal incentives for vehicle maintenance, or non-compliance fees - Fiscal incentives or non-compliance fees for driver training, especially professional drivers - Subsidies for R&D focused on vehicle performance - Annual registration taxes determined by engine size/fuel efficiency - Manufacturer fiscal incentives for performance standards (e.g. durability of fuel efficiency), or non-compliance fees	- Vehicle performance standards (e.g. durability of fuel efficiency) - Maximum power/weight ratio - Mandatory driver training, especially for professional drivers - Mandatory retirement of old vehicles - Mandatory vehicle inspection using fuel efficiency criteria - Mandatory regular vehicle maintenance - Penalties for non-compliance with standards or requirements - Speed limits (enforced)	- Automotive press coverage of vehicle performance and operating costs - Use of vehicle inspections to inform consumers about driving behaviour and vehicle maintenance - Public campaigns associating efficient driving with safe driving - Public information on efficient driving and vehicle maintenance

Table 11.2 (continued)

Policy Measures to Influence Passenger Transport Supply and Demand, and Their Areas of Impact

Area of Impact	Economic Measures	Regulatory Measures	Information
4. Vehicle use	- Fuel taxes - Insurance rates by annual kilometrage - Parking fees - Road use fees or kilometrage charges	- Car use restrictions (e.g. odd/even plate numbers) - Closing some urban districts to cars - Limits on elective or professional travel - Limits on numbers of parking spaces - Premium parking for high occupancy - Increasing congestion by not improving infrastructure - Speed limits (enforced) - Expanded/improved public transport and lanes restricted to certain users - Better infrastructure for bicycles, pedestrians)	- Public information on efficient driving and route planning - Public information on alternatives to mobility (e.g. telecommunications)
5. Fuel mix	- Fuel taxes differentiated by fuel - Taxes for high emitters/rebates for low emitters at purchase or in annual car registration fees - Fiscal incentives for development of alternative fuel distribution networks - Subsidies for R&D on alternative fuels - Rebates for conversion to dual-fuel vehicles - Low rental rates for alternative fuel equipment - Manufacturer fiscal incentives for fuel quality standards, emission standards, performance standards; or non-compliance fees	- Fuel quality standards - Emission standards - Performance standards - Penalties for non-compliance with standards or requirements	- Automotive press coverage of vehicle performance and operating costs - Use of vehicle inspections to inform drivers about alternative fuels - Education and information programmes

163

Table 11.2 (continued)

Policy Measures to Influence Passenger Transport Supply and Demand, and Their Areas of Impact

Area of Impact	Economic Measures	Regulatory Measures	Information
6. **Modal split**	- Vehicle purchase taxes - Fuel taxes - Annual licensing taxes - Annual ownership taxes - Parking fees - Road use fees of kilometrage charges - Subsidies for public transit, trains	- Car use restrictions (e.g. odd/even plates) - Ownership limits - Closing some urban districts to cars - Limits on elective/professional travel - Limits on numbers of parking spaces - Expanded/improved public transport and lanes restricted to certain users - Park-and-ride facilities - Better infrastructure for bicycles, pedestrians - Subsidies for taking alternative transport to work	- Public campaign promoting alternative transport modes - Public campaign promoting alternatives to mobility (e.g. telecommunications) - Cost comparisons for cars and public transport
7. **Fuel efficiency technology**	- Fuel taxes - Taxes for high emitters/rebates for low emitters at purchase or in annual car registration fees - Grants, loans, subsidies for fuel-efficient vehicles - Subsidies for R&D on fuel efficiency - Manufacturer fiscal incentives for fuel efficiency standards, emissions standards, power/weight ratio limits; or non-compliance fees - Tradable fuel economy credits	- Minimum fuel efficiency standards (CAFE or some other concept) - Maximum emissions standards - Maximum power to weight ratios - Penalties for non-compliance with standards or requirements	- Annual publication promoting new car fuel efficiency technology
8. **Fleet emissions abatement**	- Tax differentials favouring abatement technology on new cars - Taxes for high emitters/rebates for low emitters at purchase or in annual car registration fees - Fiscal incentives for retiring old cars - Rebates for conversion to dual-fuel vehicles - Manufacturer fiscal incentives for fuel efficiency standards, emissions standards, power/weight ratio limits, technology prescriptions; or non-compliance fees - Fiscal incentives or non-compliance fees for vehicle inspection	- Maximum emissions standards - Fuel quality standards - Maximum power to weight ratios - Technology prescriptions (e.g. catalytic converter) - Mandatory retirement of old vehicles - Mandatory vehicle inspection using emission technology and/or performance criteria - Civil penalties for non-compliance with standards or mandatory requirements	- Annual publication of new car emissions levels by fuel and vehicle type - driver awareness programmes

Intervention that goes against people's preferences generally involves significant taxes or regulations, which may distort the market and can disable domestic industry. Such distortion may be detrimental to national economies, although it is hard to obtain irrefutable evidence for this. Several OECD countries have experimented with regulatory or fiscal systems aimed at modifying consumer choices. But because consumer demand for motorisation is very strong, some policy measures have been less effective than the policy makers would wish.

Annex 9 presents the results of a study in the Netherlands of the effectiveness of a variety of measures aimed at reducing greenhouse gas emissions from transport. The study involved modelling the effects of measures on modal choices and journey routes during the peak commuter period. It does not examine effects on overall travel or vehicle energy efficiency. Within these assumptions, it indicates that significant shifts to public transport are possible by 2010. About a sixth of car users might switch to public transport either as a result of increased congestion if road capacity were not increased, or as a result of increased taxes on car use. The potential for greenhouse gas abatement in either case could exceed that from improved car efficiency.

The study also shows that measures used as proxies for direct greenhouse gas controls or taxes can have unwanted compensatory effects. Constraints on driving in residential areas can lead to more congestion in other areas and more traffic overall as drivers travel farther to avoid restricted areas. Raising parking fees in areas where parking is limited can lead to an increase in total traffic: people living close to these areas are less likely to drive, as the cost of parking is a higher proportion of their overall trip cost. This frees parking space for longer-distance travellers. Similar effects are likely elsewhere, though their extent could vary significantly. The use of combinations of measures can help by reducing this type of compensatory behaviour.

In principle, direct measures such as carbon taxes or tradable emission permits could be introduced as stand-alone policy measures. Provided a tax increment is imposed on all fuels in all sectors, in proportion to their carbon content, the result should be efficient in reducing CO_2 emissions. For many reasons, however, governments may choose other policy measures to achieve their aims. Carbon taxes can be least effective where consumers do not respond easily to price signals — for example, because there are limited opportunities to respond, or because they are unaware of the opportunities. It may also be preferable to use other policy instruments where increased taxation is not considered acceptable or tax levels are already high.

11.2.1 Internalising External Costs

External costs occur when one party's activities adversely affect other parties and no adequate compensation is made. Regulations such as emission limits can be used to prevent the generation of external costs. Alternatively, external costs can be internalised:

- by creating or enhancing related markets so that people who have to pay such costs receive compensation from those causing them — the insurance market is an example;

- through fiscal intervention to reflect the monetary value of the externalities, as with direct pollution charges;

- with proxy measures such as taxes or regulations affecting activities related to the external costs but not directly causing them.

Several recent studies aim to quantify externalities related to car use (Button, 1990; Gabel and Röller, 1992). They estimate the costs of resource depletion, congestion, accidents, pollution, noise and other consequences of car use. Their results vary widely because of differences in methodology and underlying assumptions. All estimates indicate, however, that the external costs are of the same order of magnitude as the actual monetary costs of car use. The largest, most widely recognised external costs relate to delays in congested traffic and to accidents. These may exceed air pollution costs by a factor of ten.

Evaluations of the external costs associated with climate change have been, and are being, attempted but will be of limited use until the scientific uncertainty is reduced or quantified. Hence it is not possible to offer any serious discussion of tax levels or regulations to internalise the costs of climate change. At this stage, it is more useful to examine the ways that internalising other costs would affect greenhouse gas emissions and prospects for alternative fuels.

11.2.2 Energy Security

Some OECD countries are prepared to make users pay energy prices above current world market levels to secure their energy supplies. Car use is heavily dependent on imported fuel, whose price has at times been unstable. Such instability represents a cost to national economies, which may in some cases be an externality of car use.

Most countries have taxes on transport fuel, which may help internalise the perceived security cost. Other responses include energy efficiency standards, fuel rationing and vehicle taxes related to energy use. Measures that discourage consumers from buying large cars may have unwanted side effects. If more people drive small cars, fuel economy is improved, the variable cost of driving is reduced and people may choose to drive more. In addition there is considerable public concern regarding the vulnerability of small cars in accidents.

Energy security can be affected by measures to address other externalities of car use. There is a strong connection between energy security, the promotion of alternative fuels and the reduction of greenhouse gas emissions.

11.2.3 Congestion and Traffic Management

National transport administrations traditionally view the reduction of congestion as one of their main aims. One estimate indicates that the marginal external cost of driving in peak traffic in central London amounts in 1990 to £0.36[1] per kilometre (Newberry, 1990). This is the marginal cost that each additional car imposes on all other road users in wasted time. It is about four times the monetary cost of driving a car in the United Kingdom. Government awareness of this cost has led to considerable research on ways to minimise congestion.

---

1.   £ Sterling. On average in 1992, £1 = $1.75.

Congestion affects both local pollution and greenhouse gas emissions: a car in congested traffic produces higher emissions per kilometre than a car moving at a moderate, steady speed. Congestion can be reduced in various ways:

- Road capacity can be increased, either by building roads or by improving traffic management. This results in smoother traffic flow, but this can encourage more traffic. The result is lower energy use and emissions per vehicle-kilometre, but there may be higher overall greenhouse gas emissions.

- Traffic can be discouraged by using techniques to slow speed, road user pricing, pedestrian or other exclusion zones, or by raising transport taxes. Parking restrictions, charges and fines can play an important part in reducing commuter traffic. Modal shifts can be encouraged by a combination of these techniques with public transport improvements. This type of approach generally reduces overall traffic flows, reduces emissions per vehicle-kilometre and reduces overall emissions.

- Traffic can be shifted to different roads or different times of day by strategic placement of bottlenecks, by closing some access roads at certain times, and other techniques. The overall result for the total traffic flow can be neutral, but the decrease in congestion can lead to some reduction in vehicle emissions.

Implications for Alternative Technology Vehicles. Although management of traffic flows can result in lower greenhouse gas emissions, it is unlikely to have any direct relevance to the use of alternative vehicle technology. Nevertheless, several proposals for urban traffic control projects, such as road pricing, have included suggestions to exempt low-emission vehicles. Free parking is often proposed as an incentive to promote the uptake of electric vehicles. It would be feasible to give road user fees a component related to vehicle emissions.

11.2.4 Accidents

Through the insurance market, accident costs are internalised more effectively than other externalities. Insurance companies strive to differentiate between drivers so that those who pose greater risks pay higher premiums.

Premiums will reflect the real cost of accidents only if full compensation for damages is paid by insurance companies, and it is not clear that this is always the case. In addition, important costs are associated with the risk of accidents. Costs to those who do not use cars can include stress and the time and financial costs associated with pedestrian overpasses, child car seats and other efforts to avoid accident or injury.

In addition to insurance, approaches to dealing with risk include vehicle safety standards and legislation on driving behaviour. Reduced speed limits are much discussed throughout the OECD as a mechanism for improving safety. In Europe, heavy-duty vehicles are required to have speed limiting devices. Some North American truck operators use speed limiters to reduce their fuel and maintenance costs and accident rates.

Speed limiters are not required for cars. With the increasing use of microprocessors to control engine functions it is possible to programme an engine to prevent speeding and high acceleration at low speed. These possibilities have been examined in the EC Commission's DRIVE programme. While legislation requiring such technology is unlikely in the near future, insurance companies could offer premium reductions for drivers of cars programmed for low performance. Similarly, governments might provide tax or other incentives for the use of such technologies to save energy or reduce emissions.

Despite some shortcomings, the market dealing with accident risk may have important lessons for the internalisation of other externalities. The market is founded on the recognition that those whose property or health is damaged in road accidents caused by others are entitled to compensation. The insurance market is tied to the car market in most countries through the legal requirement that all drivers should be insured.

Implications for Alternative Technology Vehicles. Measures to avoid accidents will frequently abate greenhouse gas emissions by reducing vehicle speeds. In addition, accidents are strongly correlated with vehicle performance (Kroon, 1992); full internalisation of accident costs would make high-performance vehicles less attractive and expand demand for lower-performance vehicles. This could have benefits for greenhouse gas emissions in two ways: lower-performance vehicles would be more fuel efficient and hence have lower CO_2 emissions, and the performance penalty of alternative technology would become less significant.

11.2.5 Air Pollution

The social costs of air pollution include damage to health, buildings, crops and animals (see **Annex 1**). As some local air pollutants are also greenhouse gas precursors, there is a strong connection between measures to reduce local air pollution and measures to reduce greenhouse gas emissions. Measures to reduce traffic or energy use, such as road pricing and fuel taxes, can also reduce air pollution.

Pollution Taxes. Tax systems can be designed to internalise air pollution costs. It is generally easier to relate the tax to vehicle ownership or licensing than to vehicle use. One approach is to tax car models according to the results of their certification tests. This has the advantages of using existing information and of imposing the tax when it will have most effect — when the prospective buyer chooses between car models. Such measures are proxies for taxes on actual emissions. Their main disadvantage is that the same charge is imposed on all owners of a given model, regardless of their annual kilometrage and the amount of pollution they produce. Although such taxes have not been used, some European countries have given tax relief for cars with catalytic converters in recent years. As a result, the market share of vehicles equipped with catalytic converters has increased markedly.

Vehicle emissions are measured regularly in several inspection and maintenance programmes in OECD countries. It is feasible to base pollution charges on the measured emissions from individual cars and their kilometrage between tests. One such programme is planned in Oregon in the United States. Fees averaging $50 to $200 a year are expected to result in an emission reduction of about 5% (New Fuels Report, 1992).

The low reduction in emissions expected in the Oregon plan is indicative of the very low price elasticities of car use. In addition, it is assumed that reduced kilometrage is the only response to the tax. Where alternatives exist to reducing kilometrage — for example, low-cost pollution abatement technology — these may be adopted.

Emission Standards. Fiscal intervention can be unpopular, and establishing who should pay damages and who should receive compensation can be complicated. In the case of air pollution it is hard to quantify the damage cost. It has been simpler to control air pollution through technology standards. This regulatory approach has been universally adopted in the OECD, with fiscal incentives used only as an adjunct. While standards requiring relatively cheap technical changes such as catalytic converters can be effective, standards requiring more expensive changes are harder to introduce and enforce. Governments may choose other types of measures to bring about further abatement of pollutants, including greenhouse gases.

Vehicle standards entailing the use of catalytic converters internalise the problem: they do not internalise the costs of pollution but force the vehicle owner to pay abatement costs. While this solution does not optimise resource allocation, it can result in an improvement in overall welfare, provided the aggregate abatement cost is lower than the avoided external cost of pollution.

11.2.6 Fuel Taxes

European countries generally have relatively high transport fuel taxes. This may internalise some of the external costs of car use. It also provides an opportunity for painless intervention in favour of new technology, as it is relatively simple to set up tax differentials. All countries in OECD Europe have put a higher tax on leaded gasoline than on unleaded. This tax differential, which is necessary to discourage misfuelling, also serves as a proxy air pollution tax. OECD experiences with unleaded gasoline are discussed in **Annex 8**. As **Figure 11.1** shows, a remarkably small price difference has been sufficient in the United Kingdom to bring about a very rapid switch in demand from leaded to unleaded.

This can be contrasted with France's experience introducing diesel fuel, illustrated in **Figure 11.2**. A sustained, substantial price differential between gasoline and diesel, along with other measures to promote diesel (see **Annex 7**), played a part in the growth in diesel use. The growth rate, however, was much lower than that of unleaded gasoline in the United Kingdom. This reflects the time needed for manufacturers to develop and market diesel cars, for consumers to accept them, and for the diesel cars to replace gasoline cars in the national fleet.

These two examples give some indication of the potential effectiveness of fuel taxes targeted at pollution abatement, including any that might be imposed on fuel carbon content. Unleaded gasoline is similar in performance to leaded gasoline, can be used in most existing cars without modification and can even be mixed with leaded gasoline. Its market penetration has therefore been rapid with a very small price advantage. Diesel fuel, on the other hand, can only be used in diesel cars, and diesel car buyers are making a substantial commitment, especially bearing in mind the higher vehicle cost. In this instance a lower fuel price alone is not sufficient to achieve high market shares. As **Annex 7** explains, it is important for a wide range of vehicle models to be available and for consumers to be confident that the fuel and vehicle markets will be stable.

Figure 11.1
Unleaded Gasoline Price and Share of Gasoline Market in the United Kingdom, 1985 to 1991
(%)

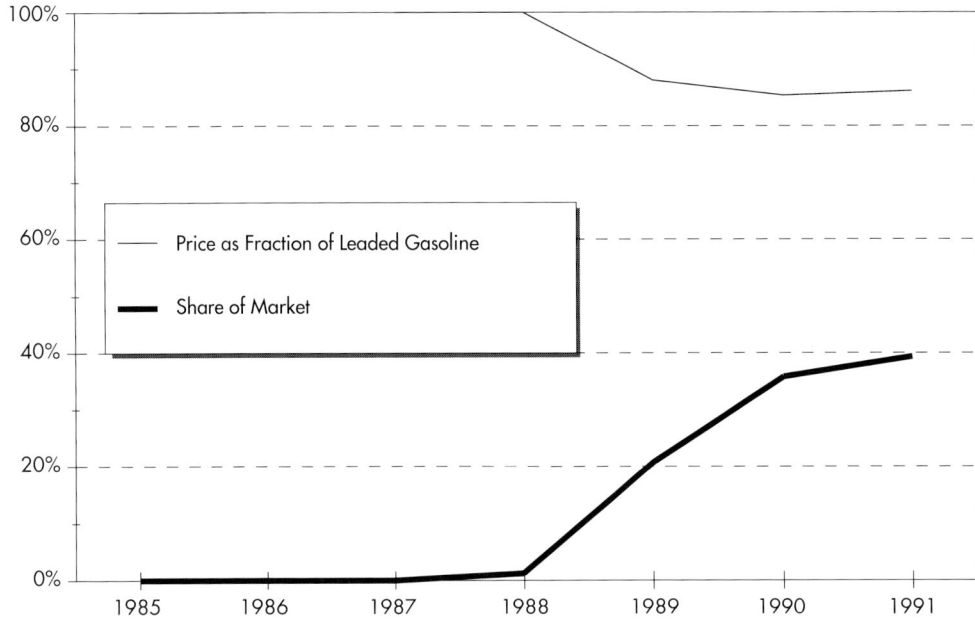

Source: Department of Transport (1992).

Figure 11.2
Diesel Fuel Price and Diesel Car Share of Fleet in France, 1970 to 1991
(%)

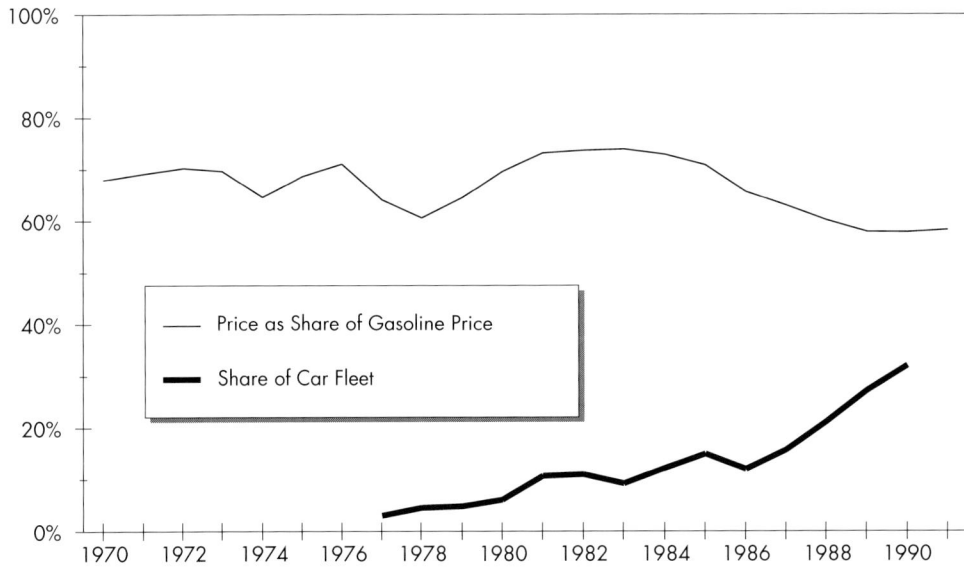

Sources: Diesel price: IEA statistics; Diesel car share of fleet: DRI European Energy Services (1989).

CHAPTER 12

CONCLUSIONS

Greenhouse Gases. The transport sector contributes a substantial and growing share of the OECD's greenhouse gas emissions. Operation of road vehicles, trains, ships and aircraft is responsible for over 30% of CO_2 from fossil fuel burning in the OECD. This figure takes account of the emissions from refineries and power stations in providing final energy for the transport sector. Vehicle manufacturing contributes a further 2-3% of the CO_2 from OECD energy use.

Road vehicles produce more non-CO_2 greenhouse gases than any other energy-using technology. Many of these gases also cause local pollution problems. It is too early to quantify the climatic effect of some of these emissions — especially NO_x, carbon monoxide and VOCs. Nevertheless, their possible importance adds to the urgency with which governments are pursuing abatement strategies.

Transport Modes. Road transport accounts for 76% of energy use in OECD transport (including marine bunkers). Gasoline vehicles — most of which are passenger cars — are responsible for 73% of road transport energy use. Bus and train use in the OECD is growing very slowly, and these modes are losing market share to cars. Of the land-based passenger transport modes, cars consume the most energy per passenger-kilometre at typical seat occupancy rates. Cars also contribute higher emissions from vehicle manufacture, as they have much lower lifetime kilometrage than buses or trains. In addition, gasoline production is generally more energy-intensive than diesel production.

Air transport's share of passenger and goods traffic is increasing. It is the most energy-intensive mode of transport, especially on short-haul flights where it competes with surface modes. Ozone precursors emitted by aircraft may have a particularly strong global warming effect. Manufacturers face a major technical challenge to maintain progress in reducing emissions and energy use by aircraft, at a time when research budgets are being cut.

Freight transport by road is increasing rapidly. This trend has serious environmental implications. Pariculate emissions could be reduced significantly by better engine maintenance, but NO_x emissions from heavy-duty diesel vehicles are harder and more costly to control than those from gasoline engine cars.

The increase in diesel fuel demand, along with increasing demand for aviation kerosene, is shifting the balance of refinery output towards premium quality middle distillates. This may make it harder to maintain fuel quality. It may also result in more energy being consumed in conversion processes in refineries.

12.1 CAR TECHNOLOGY

Changes in car technology could lead to greenhouse gas abatement. The main opportunities in the next 10 to 15 years are likely to arise from reducing energy use per unit of travel through changes in vehicle and engine design and weight, and reducing greenhouse gas emissions per unit of energy use through fuel switching. Catalytic converters will be used on most gasoline vehicles by 2005 but overall greenhouse gas emissions will not be changed significantly as a result of this technology.

About 70% of the greenhouse gas emissions from the use of an average gasoline car arise from the tailpipe and 15-20% from fuel supply. About 10% are from vehicle manufacture, depending on the life of the vehicle.

Improvements in vehicle fuel economy can result in lower emissions from the tailpipe and from the fuel supply process. Alternative fuels generally have lower tailpipe emissions but in some cases, especially for synthetic fuels such as methanol, emissions from fuel supply are likely to be higher. Emissions during vehicle manufacture become proportionally more important where fuels are derived from renewable sources. Energy use in vehicle manufacture is also likely to be higher for cars running on gaseous fuels or electric vehicles.

Fuel economy is an important element in any alternative fuel strategy. The availability of most renewable sources of fuel or electricity is limited; they would not be able to meet current transport fuel demand. Some alternatives, such as electric and hydrogen vehicles, are unlikely to be viable unless vehicles are made extremely energy-efficient.

Scope for Energy Efficiency. Manufacturers have demonstrated that it is technically possible to reduce gasoline use by cars to less than 3 litres/100 km, compared with 1990 levels of around 8-12 litres/100 km (OECD, 1991a). This potential, however, is unlikely to be achieved or even approached under the market conditions expected up to 2005. Given current market conditions, average gasoline consumption is not expected to fall much more than 10% before 2005.

Some OECD governments would like to see more rapid improvements in fuel economy. However, measures to achieve greater reductions in energy use would also be likely to affect car performance. Where governments have intervened directly to influence car design and choice they have often produced unwanted and even counterproductive results. Picking "winning" technologies is notoriously unreliable as a long term strategy. The internalisation of external costs is more promising as an approach, allowing the interplay between manufacturers and consumers in the market to determine the best technology mix.

12.2 ALTERNATIVE FUELS AND ELECTRIC VEHICLES

Alternative fuels and electric vehicles are of interest for several reasons. They have been researched and promoted in the past because of their potential to reduce dependence on oil imports, and to decrease the air pollution caused by the use of conventional fuels. The use of alternative fuels could also result in greenhouse gas abatement. The greenhouse effect due to each alternative energy source considered in this study, however, could fall in a wide range, depending on local conditions.

Many alternative fuels offer lower tailpipe emissions than conventional fuels, but higher emissions in fuel supply. Non-petroleum energy sources require a considerable amount of processing to obtain premium fuels for transport. In supplying ethanol from maize or methanol from wood, more energy is consumed in biomass production and conversion than is obtained in the final fuel product. Where process energy is supplied in the form of fossil fuels or electricity, life-cycle greenhouse emissions can be higher than those of gasoline. In comparing the environmental impact of alternative fuels, it is essential to analyse emissions from the whole process of fuel production, conversion and supply, as well as those from vehicles.

Economics of Alternative Fuels. Most renewable energy sources are currently more expensive than conventional sources. Biomass-derived fuels, including alcohol, hydrogen and bio-diesel, cost at least twice as much to produce as gasoline. They could only be marketed if they were exempt from tax and if the right vehicles were available. In the United States they can compete with gasoline only with the help of direct subsidies. Many governments consider tax exemptions and other incentives to be justified by the benefits of oil import substitution, farmer support or pollution abatement.

Some alternative fuels are cheaper than gasoline in parts of the OECD, or can be used at higher efficiency, resulting in lower vehicle operating costs. Diesel, LPG and CNG are examples; they have been used in a small number of cars in OECD Member countries for many years. Diesel and LPG have lower life-cycle greenhouse gas emissions than gasoline although CNG probably offers little reduction at present. In all three cases, the lower fuel cost is partially or totally offset by higher vehicle cost. These fuels are best suited to niches in the car market, such as fleets with high annual kilometrage. In such niches, cost considerations dominate vehicle purchase decisions and since the vehicles have high annual kilometrage the overall cost of operating alternative fuel cars is lower than that of gasoline vehicles. Diesel engines are more durable than gasoline engines. The higher lifetime kilometrage could be an important factor in reducing the overall vehicle cost per kilometre travelled. This may also apply to some extent to CNG and LPG engines. Durability is only an advantage to a car's first purchaser if the resale value reflects the car's higher life-time kilometrage capability.

Diesel vehicles make up about 10% of European new car sales. Many European car manufacturers have invested heavily in diesel engine production lines. Similar niches could be justified for LPG and CNG in North America. However, even if diesel and LPG were used by all drivers for whom they are cost-effective, greenhouse gas emissions would be reduced only slightly — probably by less than 5% of current emissions from gasoline use.

The European market for diesel cars has grown in the past 15 years, helped by diesel fuel's low price, which is due partly to a low tax rate relative to gasoline in most countries. The tax differential is likely to have macroeconomic effects that have not been studied in this report. The switch from gasoline to diesel in France involves increased consumer expenditure on domestic manufactured goods. It could result in a decrease in expenditure on imported oil, provided that the tax levels encouraging diesel use do not also result in so much additional fuel consumption as to increase imports. For a better understanding of the effective societal cost of switching to diesel fuel, it would be necessary to bring macroeconomic models to bear on this issue. It would also be important to look in detail at externalities other than greenhouse gas emissions that switching to diesel involves. A similar analysis is needed for other alternatives, especially electric vehicles.

Market Potential of Alternative Fuels. Despite the apparent financial attractiveness of some alternative fuels in certain niches, they have been slow to achieve significant market shares. There are many reasons, including lack of consumer information about the fuels and technologies, the small range of car models able to use the fuels, the limited distribution of fuel retail outlets and uncertainty regarding future fuel prices. As a result, the market potential of these fuels is low, except where governments have intervened to promote or at least facilitate their introduction.

Market entry of diesel has been encouraged not only by its price advantage over gasoline but also by its widespread availability and the commitment of manufacturers to produce and market a wide range of diesel car models. In contrast, in North America the price difference between gasoline and diesel is quite small and diesel cars are less easily available. Diesel cars are relatively rare as a result. There is considerable optimism in North America regarding the potential for alternative-fuel vehicles, including CNG, alcohol and electric vehicles. This optimism is largely a result of legislation requiring reductions in tailpipe emissions of carbon monoxide and VOCs. In the short term, however, reformulated gasoline is likely to have a major role in the response to the legislation. For the longer term, there is evidence that vehicles will be able to meet the emission standards without resorting to alternative fuels.

This study has revealed four groups of energy carriers that, in principle, can substitute for gasoline in cars:

- The first group offers life-cycle greenhouse gas emission abatement of 10% to 25% relative to gasoline. It includes diesel, LPG, CNG in optimised engines and electric vehicles operated on some existing power generation mixes. These options are commercially available in OECD countries or likely to be available by 2005. All are being actively promoted by one or more OECD governments All except electric vehicles are financially attractive for some vehicle users. The resource base for diesel and CNG is comparable with that for gasoline.

- The second group does not result in lower life-cycle greenhouse gas emissions. These fuels may be adopted for local environmental or oil security reasons. They include CNG used in existing engines; methanol from coal or gas; ethanol from high-input maize using fossil fuels for conversion; and electric vehicles using electricity from coal- or oil-fired plants. All are likely to be more expensive than current commercial vehicles and fuels.

- Offering life-cycle greenhouse gas abatement in the range of 60% to 80% are methanol, hydrogen and ethanol that is derived from wood, or other low-input biomass, with some fossil fuel input. The processes to produce these fuels have not been demonstrated on any significant scale. Costs are expected to be higher than those for petroleum-derived fuels, but lower than those for biofuels from food crops.

174

- The fourth group offers greenhouse gas abatement of 80% or more. It can include any energy carrier derived entirely from renewable sources, such as hydrogen produced by electrolysis of water using renewable electricity sources; electric vehicles powered by renewable sources; and alcohol or hydrogen from biomass without fossil fuel inputs. Many of these carriers require considerable technical development before they can be widely used. The complete elimination of fossil fuels from biomass production and processing would probably add to the costs.

12.3 POLICIES AND STRATEGIES

There are many non-monetised costs and benefits associated with the use of cars. They are of great significance both to vehicle owners and to the rest of society. Car ownership and use are extremely important to consumers and to OECD economies. Attempts by governments to limit the number or size of cars on the roads have tended to be ineffective. Implementing policy measures that restrict access to such a highly valued commodity is likely to be difficult and costly.

There is considerable experience in the OECD with policies aimed at reducing traffic levels or transport energy use. Many of these have been aimed at proxies. For example, energy efficiency standards have been used to try to reduce energy use, and road access restrictions have been used to try to reduce traffic. Such proxy measures often have unwanted side effects. Improved vehicle efficiency makes travel cheaper and can actually increase energy use, although it may deny consumers the level of performance, comfort and safety they would otherwise have chosen. Road access restrictions often result in increased traffic overall as drivers take longer routes to avoid restricted zones. Proxy measures for greenhouse gas abatement can only be fully effective if comprehensive packages are designed to prevent consumers from taking compensating action.

Direct greenhouse gas abatement measures would be more efficient but might be unacceptable. Taxes related directly to the carbon content of fuel are one example. Such taxes would be most efficient if applied world-wide to all energy-using sectors. In several OECD countries transport fuels are already heavily taxed. Carbon taxes at the levels that so far have been discussed or applied would have less impact on energy use in transport than in other sectors.

There is growing concern about other effects of transport. Congestion, accidents and local air pollution are widely recognised as issues requiring urgent action. Perhaps the most fruitful approach is to consider greenhouse gas abatement alongside these issues as an additional externality to be taken into account. Measures to reduce these problems include transport demand management, speed limit enforcement and incentives to reduce vehicle power/weight ratios.

Government responses to problems associated with transport have generally been fragmented, partly because of the institutional history of intervention in the transport sector. Energy security may be managed by trade or energy departments, fuel taxation by finance ministries, congestion by transport departments, air pollution by environment departments, accidents by health and police administrations. In view of the interaction among various policy issues, it is important to consider all effects of car use together. Even where external costs cannot be monetised, an integrated approach is essential to the formulation of well-balanced, effective policy strategies.

Technical solutions to the problems caused by transport differ according to local priorities. For example, electric vehicles can result in lower urban pollution, but they will not reduce congestion and they may result in increased energy use and greenhouse gas emissions. In regions with fossil-based power stations and limited sulphur and NO_x controls, emissions of acid deposition precursors could be increased by moving to electric vehicles. Similarly, alternative fuels such as CNG used in converted gasoline cars and methanol from natural gas can be used to reduce oil dependence but will not reduce greenhouse gas emissions and may not affect local pollution. National analyses are needed to identify which technologies and policies best satisfy local needs.

At the same time, in a general environment of increased internationalisation of the economy, there is scope for co-ordination of strategies between OECD member countries and between governments and industry. A small number of significant car manufacturers is concentrated in an even smaller number of countries. While there is a convergent tendency in car and engine design, manufacturers produce cars to a variety of specifications, to meet emission standards which differ between countries. The multiplicity of emission standards raises manufacturers' costs of competing in the international market. The case may be strengthening for establishing common vehicle emission standards where possible. This would be particularly important for alternative fuel vehicles where the market in any one country is too small to justify mass-production.

Many governments aim to reduce the negative effects of transport by internalising the societal costs associated with car use and by making sure that transport users have sufficient information to act on the cost signals they receive. Some externalities can be internalised through the creation of markets; accident costs are internalised to some extent through the insurance market. Other costs can be internalised through fuel taxes, vehicle purchase taxes and licensing fees, and road user charges. In using a taxation approach it is necessary first to quantify the external costs. This is an important area for further study. In cases where it is difficult to ensure that consumers are adequately informed, it can be more effective to introduce regulations and standards. Such measures have been the preferred approach for reducing pollution from vehicles.

Policy Effects on Greenhouse Gas Emissions. Measures to reduce accidents, congestion and air pollution will tend to reduce energy use and greenhouse gas emissions, while the need for greenhouse gas abatement adds to the pressure to deal with these other problems and could be the deciding factor for some governments in taking action.

In some cases, internalising external costs will encourage energy efficiency and the use of alternative fuels, which have particularly strong interactions with security of supply and air pollution. Internalising other external costs, such as those of accidents and congestion, will probably have little effect on the attractiveness of fuel switching.

Even if known external costs are fully internalised, alternative fuels are unlikely to be attractive enough to overcome most consumers' preferences for well-known, convenient technology. Most use of most alternative fuels, therefore, is likely to continue to be in niche markets such as commercial and government fleets.

The introduction of alternative fuels requiring specialised vehicles is likely to be a slow process, and it may be expensive. Consumers are only likely to buy such vehicles if they are sure that the fuel price will remain attractive. The goal is most easily achieved where fuel or vehicle suppliers are willing and able to arrange for the financing of vehicles and fuel supply infrastructure.

Alternative fuels that can be mixed with conventional fuels in existing vehicles are much easier to introduce and do not require long-term commitment. Governments generally have used the promotion of such fuels to fulfil other short-term policy objectives — in agriculture, for example. Existing fuels of this type do not bring significant life-cycle greenhouse gas emission benefits, though local air pollution may be reduced.

12.4 FURTHER WORK AND RECOMMENDATIONS

Although this report has demonstrated the use of life-cycle analysis for comparing technologies, many assumptions are made for purposes of illustration only and the results of the analysis should not be used for national decision making. Understanding of the issues addressed in this report will be improved substantially if governments co-operate in carrying out their own analyses and sharing the results.

The study has focused on the life-cycle greenhouse gas emissions of alternative-fuel cars and their cost to the user. Further analysis is needed to quantify the external costs and benefits of such technology over the full life-cycle. Analysis is also needed of the energy use and emissions associated with car disposal, infrastructure provision, the transport service industry, police operations and other activities associated with car use. The energy and environmental implications of the trend toward greater recycling of car components need to be understood.

Macroeconomic studies are needed for a better understanding of the effects of using more expensive technology to reduce oil consumption. For example, it is not clear whether manufacturing industry output and employment would rise because of the higher added value per vehicle, or fall because of reduced demand for cars.

The fastest-growing transport subsectors are air travel and road freight. There is considerable need for closer analysis of the technical, economic and socio-political issues raised in these sectors.

International studies such as this report are hampered by the poor availability and comparability of transport data. Most countries have no definitive statistics on energy used or distances travelled by particular classes of vehicles. This makes it impossible to monitor developments in energy efficiency with accuracy and hence to assess the effectiveness of policy measures. If such databases were to be developed, it would become easier to carry out more detailed econometric studies of policy options for transport. This report urges governments to improve the quality of their statistics as a fundamental basis for effective policy making.

ANNEXES

ANNEX 1

AIR POLLUTION FROM TRANSPORT

This annex presents some of the effects of transport air pollution. It then discusses the state of knowledge regarding global warming.

Emissions from transport have many types of effects. They are summarised in **Figure A1.1**. Welfare effects include consequences for economic production, ecosystems and infrastructure, summarised in **Table A1.1**. Health effects are presented in **Table A1.2**.

Figure A1.1
Emissions from Transport: Local, Regional and Global Effects

IMPACT	Particulate Matter	Lead	Sulphur Dioxide	NO_x	VOC	Carbon Monoxide	Methane	CO_2	N_2O	CFCs
Local (Health and Welfare)	X	X	X	X	X	X				
Regional										
Acidification			X	X						
Photochemical Oxidants				X	X	X				
Global										
Indirect Greenhouse Effect				X	X	X	X			X
Direct Greenhouse Effect							X	X	X	X
Stratospheric Ozone Depletion				X					X	X

Source: Faiz, A. (1991).

Local and Regional Effects

Acid Deposition. Acid deposition results mainly from emissions of oxides of sulphur and nitrogen. Dry deposition occurs close to the source of the emissions. Wet deposition can occur over much longer ranges, in excess of 1 000 km, especially where emissions are produced from tall stacks to avoid local pollution.

Table A1.1
Welfare Effects of Pollutants from Motor Vehicles

Pollutant	Welfare Effects
Carbon Monoxide	No adverse effects on vegetation at commonly observed concentrations.
NO_x, Sulphur Dioxide	Precursors of acid deposition (acid rain), deposited as dust or precipitation: damage to aquatic systems, fresh-water fisheries and forest ecosystems, leaching of sensitive soils. Corrosion of metals and other materials; structural damage to ornamental stone facades, buildings and monuments; degradation of nylon and cotton textiles. NO_x causes a brown pollutant haze, reducing visibility. Cause of groundwater pollution by heavy metal leachates.
Ozone, Other Photochemical Oxidants	Damage to forest ecosystems particularly from secondary stress factors. Even low concentrations can cause damage and impair growth of a variety of cash and food crops (e.g. wheat, corn, soyabeans, peanuts), vegetables and fruits. Causes premature cracking of rubber products. Contributes to fading of nylon carpets and polyester-cotton fabrics in high humidity. Degrades visibility, affecting air and surface transportation.
Hydrocarbons	Ethylene is one of the few hydrocarbons from raw exhaust that damages fruits and plants.
Particulates and Smoke	Diesel smoke and particulate emissions can significantly reduce visibility; also associated with soiling of buildings, materials and fabrics. Source of unpleasant odours. Deposits on vegetation may inhibit photosynthesis.
Lead	Lead accumulates in plant and animal tissue, which may be toxic if consumed as food.

Source: Faiz (1991)

Table A1.2
Health Effects of Pollutants from Motor Vehicles

Pollutant	Health Effects
Carbon Monoxide	Interferes with absorption of oxygen by red blood cells; impairs perception and thinking, slows reflexes, causes drowsiness, brings on angina and can cause unconsciousness and death. It affects foetal growth in pregnant women and tissue development of young children. It has a synergistic action with other pollutants to promote morbidity in people with respiratory or circulatory problems; associated with reduced worker productivity and general discomfort.
NO_x	Can increase susceptibility to viral infections such as influenza; irritate the lungs and cause oedema, bronchitis and pneumonia; and result in increased sensitivity to dust and pollen in asthmatics. Most serious in combination with other pollutants.
Hydrocarbons	Low molecular weight compounds cause unpleasant effects such as eye irritation, coughing and sneezing, drowsiness and symptoms akin to drunkenness; high molecular weight compounds may have carcinogenic or mutagenic effects. Some hydrocarbons carried on diesel particulates may cause lung disease.
Ozone (Precursors: Hydrocarbons and NO_x)	Irritates mucous membranes of respiratory system, causing coughing, choking and impaired lung function; causes eye irritation, headaches and physical discomfort; reduces resistance to colds and pneumonia; can aggravate chronic heart disease, asthma, bronchitis and emphysema.
Lead	Affects circulatory, reproductive, nervous and kidney systems; a suspected cause of hyperactivity and lowered learning ability in children. Hazardous even after exposure ends. Lead is ingested through the lungs and the gastrointestinal tract.

Health Effects of Pollutants from Motor Vehicles

Pollutant	Health Effects
Sulphur Dioxide	A harsh irritant; exacerbates asthma, bronchitis and emphysema; causes coughing and impaired lung functions.
Particulate Matter	Irritates mucous membranes and may initiate a variety of respiratory diseases; fine particles may cause lung cancer and exacerbate morbidity and mortality from respiratory dysfunctions. A strong correlation exists between suspended particulates and infant mortality in urban areas. Suspended particulates can adhere to carcinogens emitted by motor vehicles.
Toxic Substances	Suspected of causing cancer, reproductive problems and birth defects. Benzene and asbestos are known carcinogens linked to leukaemia and lung cancer; aldehydes and ketones irritate the eyes, cause short-term respiratory and skin irritation and may be carcinogenic.

Note: There is growing evidence that the synergistic effects of these pollutants may be far more serious than the effects of individual pollutants. This is particularly the case where NO_x and SO_x and certain hydrocarbons co-exist or occur in association with particulate matter.

Source: Faiz (1991).

Photochemical Oxidants. A large number of pollutants react with each other and with gases in the atmosphere, in the presence of sunlight, to produce secondary pollutants. NO_x and hydrocarbons are involved in these reactions. The pollutants produced include ozone, peroxyacetyl nitrate and hydrogen peroxide. Raised ozone levels can be observed some distance from the original pollution, several days later, depending on air movements. Emissions of a pollutant in one region can react with emissions of another pollutant in a different region. For example, urban emissions of NO_x can mix with VOCs from forests to result in high rural ozone levels.

Global Warming

The "greenhouse effect" is a well-established phenomenon, without which the temperature at the earth's surface would be about -30°C: at this temperature the thermal radiation from the earth's surface would exactly balance the incoming radiation from the sun.

"Greenhouse" gases, mainly water vapour and CO_2, are transparent to the visible light that forms much of the incoming solar radiation, but they absorb the infrared radiation produced by the earth's surface. Although the gases re-radiate the energy, the temperature of the troposphere rises closer to that at the earth's surface. The eventual result is warming at the earth's surface.

An anthropogenic, enhanced greenhouse effect may be occurring, caused by emissions of a number of gases. Some have a direct radiative impact — CO_2, methane, N_2O and chlorofluorocarbons (CFCs) being the most important — and some react in the atmosphere to affect concentrations of greenhouse gases, especially methane and ozone. CFC replacements such as HFC 134a may make important contributions to the greenhouse effect.

The considerable literature on the sources, sinks and atmospheric concentrations of greenhouse gases is summarised in publications by the Intergovernmental Panel on Climate Change (IPCC, 1990 and 1992).

In order to understand the possible relative effects of technologies on global warming, it is helpful to have a common basis on which to compare different gaseous emissions, including emissions of indirect greenhouse gases. There is no generally agreed methodology for making comparisons. Greenhouse gases vary in their radiative effect for a given atmospheric concentration and in the time emissions remain in the atmosphere.

One way to compare emissions is through their "global warming potentials" (GWPs). The GWP of a gas is a weighting factor that indicates its importance as a greenhouse gas relative to a reference gas, normally CO_2. GWPs are generally calculated on a mass basis. So as to deal with the different lifetimes of the gases, GWPs indicate their warming effect integrated over a fixed period or "time horizon".

The GWP of a gas is defined to be the warming effect, integrated over the time horizon, of an emission of a kilogram of the gas, divided by the warming effect over the same period of a kilogram of CO_2. The approach is accurate only if:

- the gas is well-mixed in the atmosphere;

- both CO_2 and the other gas have well-known variations of concentration over time;

- effects of adding more of the gas to the atmosphere are linear and effects of different gases are independent.

The assumption of good mixing applies to long-lived gases such as CO_2, N_2O and methane. It does not apply to short-lived gases, especially the ozone produced by NO_x, carbon monoxide and VOCs.

Variations in concentration over time are being researched and frequently revised. The carbon cycle itself, which determines the variation in CO_2 concentration, is poorly understood. The behaviour of many gases depends strongly on the level of water vapour and the hydroxyl (OH^-) radical in the atmosphere, these in turn depending on pollution levels. Variations in global NO_x and hydrocarbon emissions of around 10% appear to be approximately linearly independent (Hough and Johnson, 1990).

The GWPs used in this study are shown in **Table 2.1**. They are based on figures originally published by the IPCC (1990) as examples of the use of GWPs. The GWPs for NO_x, hydrocarbons including methane, and carbon monoxide are based on complex calculations using models of atmospheric physics and chemistry. After the publication of the IPCC's 1990 report the figures were widely used, but later it was shown (Johnson *et al.*, 1992) that they contained an error in the effects due to ozone formation. They also omitted effects due to changes in the methane concentration: methane levels are increased by emissions of carbon monoxide and hydrocarbons but reduced by emissions of NO_x. A considerable amount of additional work is needed to improve understanding of these indirect effects.

It is uncertain whether the GWP of NO_x is positive or negative, as there is a positive impact on ozone concentrations and a negative impact on methane concentrations. For calculations in this report, the GWP of NO_x is taken to be zero.

In the rest of this study emissions are calculated on a life-cycle basis. Technology emission factors have been identified for a broader range of greenhouse gases, including VOCs and N_2O. CFCs are not considered: they are emitted not as a direct result of energy use but rather because of leakage and venting from air conditioning systems. Their direct greenhouse impact is very large, but it may be offset by a reduction in radiative forcing due to ozone in the upper troposphere, which is depleted by CFCs. They are expected to be phased out in the transport sector during the 1990s as a result of the Montreal Protocol of 1988.

POTENTIAL CO$_2$ IMPACT OF ALPTRANSIT RAIL PROJECT[1]

To accommodate growth in freight traffic to 2010 and beyond, the Swiss Government has proposed a modern transalpine rail system, "AlpTransit", crossing the country from Italy to Germany. While the full system will not be finished until 2010, roughly half of the expansion required is expected to be in place by 1994.

AlpTransit's objectives are to:

- develop an alternative to increasing road freight transport between northern Europe and Italy;
- integrate Swiss passenger rail transport into the evolving European network of high-speed trains.

The system will be electric. The share of fossil fuels in Swiss power generation is small both on average and at the margin, so greenhouse gas emissions due to train operation will be low.

Most freight between northern Europe and Italy is now taken through Austria and France, even though the most direct route is through Switzerland. Swiss regulations limit gross vehicle weights to 28 metric tons and prohibit Sunday and night driving. Consequently the share of road traffic in freight transport in Switzerland is less than one-third that in the neighbouring countries. Under an agreement with the European Community, these limitations are dropped when rail capacity is saturated.

AlpTransit and Strategies to Reduce Greenhouse Gas Emissions

A study of the effects of AlpTransit found that it would result in less use of road transport fuels and hence reduced greenhouse gas emissions. This will be slightly offset by increased fuel consumption for power generation. The net emission balance depends on how the additional electricity is produced and the load factors that can be achieved on the rail system relative to that of the road traffic displaced.

---

1. Based on a report to the IEA by the Swiss Government.

Table A2.1
Expected Impact of AlpTransit Project in 2010

		Absolute Values		Capacity Impact, 2010	
	1990 Effective	Capacity, 2010		Absolute Change	As a percentage of Reference
		Without AlpTransit (Reference)	With AlpTransit		
1. Annual passengers & freight[1]					
Passengers (million)	7	11	21	+ 10	+ 91%
Freight (million metric tons)	13	30	69	+ 39	+130%
2. Trains per day					
Passenger trains	110[1]	130	150	+ 20	+ 15%
Cargo traffic	150	280	400	+120	+ 43%
Total	260	410	550	+140	+ 34%
3. Road Transit and Traffic[2]					
Cars/day	13 000	20 000	18 500	-1 500	- 8%
Trucks/day	2 100	3 500	1 000	-2 500	- 71%

1. Swiss alpine rail traffic for Gotthard and Lötschberg without local trains.
2. Traffic on National Highway N2 from Basel to Chiasso.

AlpTransit will involve:

- developing a combined road-rail system for long-distance freight;
- reducing time losses at national borders;
- internalising the external costs for all transport modes;
- halting capacity expansion in the transalpine road system;
- maintaining the truck weight limit and the ban on night and Sunday driving.

The financial viability of the project depends on high load factors at adequate prices. The strategy will involve making rail transport more attractive and road transport less attractive.

The switch from road to rail should reduce the number of cars and trucks on Swiss roads and on the Brenner highway. Rail traffic that currently uses the Brenner rail system will to some extent be transferred to the AlpTransit system. This in turn will make capacity available on the Brenner rail system, which can then absorb and further decrease road transport on the Brenner highway. Re-routing away from the Brenner systems will reduce distances travelled by an average of 100 km per trip.

The expected effects of AlpTransit on traffic are shown in **Table A2.1**. **Table A2.2** shows the effect on energy use and CO_2 emissions in the first decade of full operation.

Table A2.2
**Summary of AlpTransit's Impact on Electricity Demand,
Fuel Savings and CO_2 Emissions, 2010-2020**

	Within Switzerland	Total System
Annual electricity demand (GWh)	500-600	800-1 000
Annual fuel savings (million litres)	70-100	500-800
CO_2 emissions due to AlpTransit (1 000 metric tons/year):		
From electricity production	+60	+100
From fuel savings	-200	-1 500
Net savings	-140	-1 400

ANNEX 3

CO$_2$ PERSPECTIVES IN SWISS AIR TRAFFIC[1]

Introduction

This case study focused on Swiss air traffic in an effort to identify means of reducing CO$_2$ emissions. The study found that allocation of aviation fuel use to countries is not straightforward. Several approaches are possible:

- Aviation fuel consumed within national borders gives one indication. Swiss aviation fuel consumption results in the emission of roughly 3 million metric tons of CO$_2$. Much of the fuel is consumed by foreign airlines carrying citizens of other countries.

- Flights over the territory give another indication. CO$_2$ emissions related to air traffic over Switzerland are estimated to amount to 1 million metric tons. This approach carries a risk of underestimation, however, as it does not allocate flights over oceans to any country.

- Fuel consumption by air traffic to and from Switzerland is estimated to result in 4 million metric tons of CO$_2$ emissions. This measure, applied globally, double-counts the amount of fuel used; the actual allocation to Switzerland would be 2 million metric tons.

Prognoses for Fuel Demand and Efficiency

Forecasts indicate that, by 2005, air passenger traffic in Switzerland will have increased by 220%. Factors influencing the forecasts include global and regional economic development; institutional frameworks; the availability and attractiveness of alternative transport modes, especially rail; airport capacity limits; and political priorities.

Fuel consumption per passenger-kilometre in the global aviation industry declined by 3.5% a year from 1960 to 1990. Key factors in this decline include engine and aerodynamics improvements, reduced weight through better materials, increased aircraft size and bigger load factors. A reduction of 30% in fuel consumption per passenger-kilometre is expected to be possible by 2010 and 40% by 2025, based on the following factors:

- New propulsion technologies, such as very high bypass-ratio engines, may reduce energy use by 15%.

---

1. Based on a consultant's report provided to the IEA by the Swiss Government.

- Better aerodynamics through such advances as improved aerofoil design and laminar flow technology, and reduced weight through the use of graphite, fibre-plastic composites and other materials, could offer fuel savings of 10-15%.

- Operational factors such as trim control, improved route planning and better engine and instrument maintenance could save 3-5%.

- Improved passenger and freight load factors could reduce energy use by 5-10%.

The effect of the 220% increase in traffic on fuel demand is expected to be offset by improved energy efficiency, resulting in an increase in Swiss jet fuel consumption of roughly 50% by 2005. The study expects fuel consumption per passenger-kilometre to improve more rapidly after 2005.

Potential Measures to Reduce CO_2 Emissions

Price Policy. The case study identifies taxation as an important issue. Aircraft fuel is exempt from taxes world-wide, in accordance with the International Civic Aviation Organisation's Resolution A26-15 ("Imposition dans le domaine du transport aérien international"). The resolution gives air transport preferential treatment relative to other modes in most countries. Aircraft fuel taxes cannot be imposed unless the resolution were to be revoked, which would be difficult. Given the global nature of air traffic, fuel taxation must be internationally co-ordinated, according to the case study. A tax on air transport could be imposed directly on the fuel used, or collected as landing fees and flight route fees related to CO_2 emissions[1]. A CO_2 tax on aircraft fuel could have several effects:

- encourage fuel efficiency;

- decrease demand for air transport, especially in short-range leisure travel, where alternative modes are available: high-speed rail and cars can compete with air travel for trips of up to 1 000 km. Long- and medium-range traffic and short-range business travellers are less likely to switch modes but may be affected to some extent;

- raise revenue for energy and environment R&D.

Policy Options. The main instruments for reducing CO_2 emissions from air transport are:

- subsidising R&D;

- imposing energy efficiency standards on manufacturers, combined with civil penalties, non-compliance fees or fiscal incentives;

- halting airport capacity expansion;

- encouraging switches to rail, especially in short-range air traffic.

Using bilateral air traffic agreements and licences, restrictions could be placed on the number of flights and on fuel consumption. Such measures would have to be imposed according to generally accepted criteria, such as aircraft load factors and potential for substitution. The impact of the measures would depend on the number of countries adopting restrictions.

---

1. Such fees could take account of fuel consumption per passenger-kilometre and distance travelled, and exemptions could be provided for fuel-efficient technology.

Assessment of Measures

The case study found that a tax on aircraft fuel would not have a significant impact on air transport demand unless it was fairly high, and in that case it would have to be introduced gradually and co-ordinated internationally. It could be introduced as a complement or an alternative to supply restrictions. Landing fees could be introduced by individual countries.

Supply restrictions for medium- and long-range traffic could be effective in the medium-term but politically difficult to impose. Such measures strongly encourage higher load factors and the use of larger aircraft, improving energy efficiency.

Taxes and supply restrictions would affect aircraft manufacturers by reducing demand for aeroplanes and encouraging development and uptake of new technology. Airlines and both upstream and downstream industries would also be affected. Hence it might be politically advantageous to implement such measures gradually, starting with those that would have a relatively mild impact on economic variables.

ANNEX 4

NORWEGIAN COASTAL SHIPPING

The Norwegian Government commissioned a study for the IEA of emissions from ships in Norwegian coastal waters (Marintek, 1991). The study examines data on energy use and emissions from various types of shipping. It considers the outlook for emissions to 2005 and examines technical means and policy measures that could reduce emissions.

Marine transport is responsible for most of the goods traffic and about a quarter of transport energy use in Norway, as **Figures A4.1** and **A4.2** show. Of the bunker fuel used in Norway's 200 nautical mile economic zone, over 60% is consumed within 12 nautical miles of the coast. The most significant elements of marine energy use are coastwise shipping, domestic ferries, international trade and fishing.

Figure A4.1
Domestic Goods Transport, Norway, 1970 to 1990
(metric ton-km per capita)

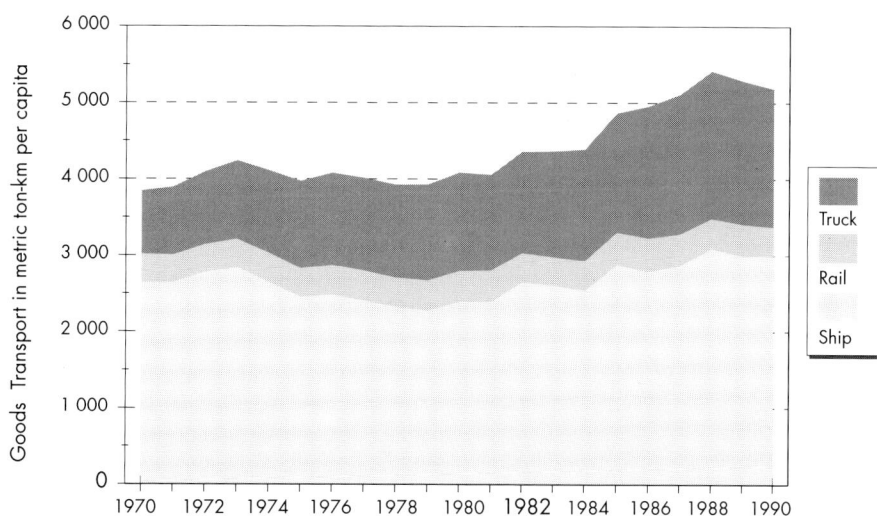

Source: Schipper (1992).

195

Figure A4.2
Energy Use in Norwegian Transport, 1988
(PJ)

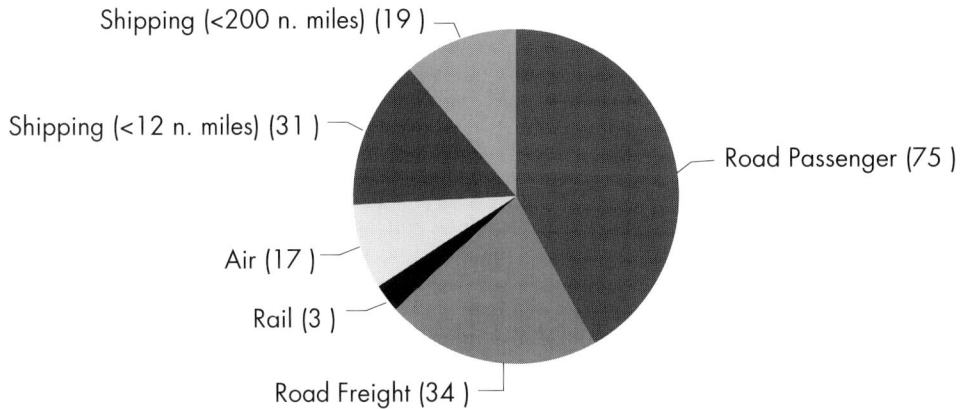

Shipping (<200 n. miles) (19)

Shipping (<12 n. miles) (31)

Air (17)

Rail (3)

Road Freight (34)

Road Passenger (75)

Source: Shipping energy use: Marintek (1991); Other modes: Schipper (1992)

Because of a lack of documentation regarding the use of bunker fuel, it is difficult to define the amount of fuel used by Norwegian shipping. Ships operating under the Norwegian flag refuel in foreign ports, and ships operating under other flags refuel in Norwegian ports. Some domestic waterborne transport may use international marine bunker fuel. There are considerable differences in the shipping energy use reported by various agencies. The IEA figures for internal navigation are nearly 30% lower than those of the Norwegian Central Bureau of Statistics.

The share of shipping in surface goods traffic is decreasing, after a particularly large rise in road freight during the mid-1980s. The study forecasts constant bunker fuel demand in the period to 2005. CO_2 emissions can be reduced by fuel switching; the report considers the effect of switching to CNG and LNG. In the scenario considered, natural gas is used to fuel 2% of coastal ferries by 1995. Of ships and ferries in coastal trade, 30% switch to natural gas by 2000 and 40% by 2005.

Emissions of sulphur and NO_x from ships are of particular concern. Sulphur emissions are especially high where heavy fuel oil is used. Ship flue gases can be scrubbed using sea water to remove acid precursors. In the scenario the technique is used on 95% of oil-fuelled ships by 2005. All other oil-fuelled ships operate on marine gas oil with a maximum sulphur content of 0.05%.

NO_x emissions can be reduced by retarding fuel injection in engines, by using exhaust gas recirculation, or through selective catalytic reduction (see **Chapter 8** for further discussion of these techniques). The report foresees extensive use of retarded injection in marine engines to give emission reductions of up to 50%, with a 5-10% fuel consumption increase.

The overall effect in the scenario is a reduction in emissions as shown in **Table A4.1**. These reductions are compared with targets proposed by the International Maritime Organisation in 1990: a 50% reduction in sulphur dioxide and a 30% reduction in NO_x emitted by ships by 2000.

Table A4.1
Emission Outlook for Norwegian Coastal Shipping

Gas	Outlook			
	1988	1995	2000	2005
SO_2 (thousand metric tons)	29	24	14	4
NO_x (thousand metric tons)	82	80	69	51
CO_2 (thousand metric tons of carbon)	1 000	999	979	971

ANNEX 5

EMISSIONS MODEL

This study uses a life-cycle greenhouse gas emission model to calculate emissions associated with the use of vehicles[1]. The model takes the form of several linked spreadsheets. They account for emissions in vehicle use and manufacture and the energy supply system. The model calculates the life-cycle emissions per unit of distance driven by each of several vehicle types and using several fuel sources.

The spreadsheets calculate the energy inputs to each stage in the chain of fuel conversion, transport and distribution. **Table 6.2** shows the energy use in each stage for the various fuels. **Table A5.1** shows vehicle energy use for alternative fuels and gasoline.

Table A5.1
Car Weight and Energy Use Assumptions

Vehicle Type	Fuel Use (MJ/km)	Driving Range (km)	Storage Sys. Weight (kg/kg fuel)	Fuel Weight (kg)	Vehicle Weight (kg)
Gasoline	2.30	560	0.4	27.5	860
Diesel	2.04	710	0.4	31	1 028
LPG	2.12	560	1.33	31.4	877
CNG	2.16	400	4.39	16.2	909
Methanol	2.00	560	0.31	49.4	860
Ethanol	2.02	560	0.33	37.9	860
Liquid Hydrogen	1.86	560	5.76	7.4	871
Electric Vehicle	0.69	210	—	365.6	1 109

Energy inputs are broken down by fuel, and emissions from fuel combustion are calculated. Emission factors are used to calculate non-CO_2 emissions from fuel use. They are specific to the processes and fuels. Most are derived from the EPA database published as AP-42 (EPA, 1985 and 1988).

---

1. The model was produced by Dr. Mark A. DeLuchi of the Institute of Transportation Studies, University of California, Davis, California. It formed the basis of his PhD dissertation.

The one exception to this description is for a plant converting natural gas to methanol: here CO_2 emissions are calculated as the difference between carbon in the feedstock and carbon in the output rather than as the product of an emission factor and a process-energy consumption figure. However, emissions of non-CO_2 greenhouse gases from the plant, as well as off-site emissions from the generation of electricity consumed by the plant, are calculated as above.

Sources of Energy Data. Data from many sources are used to calculate the amount and kind of energy used at each stage of fuel production and use. Where possible, historical aggregate data are used. They are available for coal and natural gas in some OECD countries. For these industries it is possible to calculate the process energy use per unit of fuel for final consumption.

Data on other fuel supply processes are either unavailable or less tractable for the analysis carried out here. In the cases of oil products and nuclear power, aggregate data exist on process energy use at some stages in the supply process. The processes and outputs are more complex than those for natural gas and coal, however, and it is not possible to calculate directly the energy use at all fuel processing stages per unit of fuel for end-use. Estimates have therefore been used.

Several of the alternative fuels considered are not in widespread production and little if any historical information is available on energy use in their supply processes. It has therefore been necessary to rely on estimates of energy use from engineering studies and monitoring of prototype plant.

For fuels other than coal and natural gas, the model calculates the amount of process energy used to produce one unit of energy available to the transportation sector. Use of the product fuel in further processing and distribution is accounted for as "own use".

A careful analysis has been made of "own use" in oil product distribution. Diesel tankers sometimes obtain fuel at service stations while at other times they take from storage terminals fuel that has not been through the final stage in distribution. In the model, however, all fuel used by tankers is assumed to have been distributed to filling stations. This assumption has a negligible effect on the results of the model (less than 0.01% possible error).

Second-Order Effects. The model includes calculation of the emissions involved in producing fuel used in energy conversion; e.g. coal and gas used at oil refineries. Oil use at refineries is treated as "own use" and upstream emissions do not have to be separately accounted for as they are implicitly counted through the reduced output of the fuel chain.

Circular Effects. Several fuel supply chains are interdependent. For example, refineries use electricity and refinery products are used in power generation; i.e. some oil products are used to produce electricity that is then used to produce more oil products. In calculating emissions from the supply of oil products, it is necessary to account for exchanges of this type.

In principle this use of oil could be treated as "own use" by refineries. In practice, however, the calculation would be complicated. Instead, an iterative approach has been adopted. The upstream emission factors calculated for oil in one model run are applied to the oil used in power stations in the next model run.

Vehicle Energy Use. Vehicle energy use is calculated first for gasoline and diesel vehicles from fuel economy data entered by the model user. Energy use is calculated for the alternative-fuel vehicles based on assumptions about their drive-train efficiency relative to that of a gasoline vehicle. The effect of vehicle mass on energy use is also taken into account, using a linear relation that depends on the driving cycle.

Non-Combustion Greenhouse Gas Emissions. The model includes detailed calculations of venting, flaring or leaking of gas — mostly methane — from coal beds, oil wells and natural gas systems. It also incorporates emissions from the manufacture and assembly of materials used to make vehicles. Consideration has been given to energy use in the construction of conversion plant, such as power stations and oil refineries, but this has not been integrated into the model. In addition to energy use for fuel transport, estimates have been made of other energy use associated with fuel transport and distribution — in truck maintenance, for example.

ANNEX 6

SENSITIVITY ANALYSIS FOR COST CALCULATIONS

This annex provides the results of sensitivity analysis, carried out for the cost calculations in **Chapter 7**. **Tables A6.1** and **A6.2** present the results of varying some of the assumptions on which the calculations are based.

Table A6.1
Cost of Fuel Switching in the United States
Sensitivity Analysis Results
Base case: Travel 16 400 km/year; 5% Discount Rate; All other data as in **Table 7.2**

Case		Diesel			CNG 2/3 Range			CNG Full Range		
		High	Low	Midpoint	High	Low	Midpoint	High	Low	Midpoint
		Cost of Fuel Switching in US cents per km								
Base	Total	2.63	0.68	1.65	0.39	-1.19	-0.40	1.44	-0.62	0.41
	Ex Tax	2.60	0.74	1.67	0.37	-1.17	-0.40	1.35	-0.67	0.34
High Annual Travel 25 000 km/year	Total	-0.08	-1.49	-0.79	0.10	-1.32	-0.61	0.98	-0.84	0.07
	Ex Tax	0.03	-1.32	-0.64	0.10	-1.30	-0.60	0.91	-0.87	0.02
Low Annual Travel 12 000 km/year	Total	3.42	0.89	2.16	0.91	-0.91	0.00	2.25	-0.21	1.02
	Ex Tax	3.36	0.93	2.14	0.87	-0.91	-0.02	2.12	-0.27	0.92
High Discount Rate 30%	Total	5.21	1.56	3.39	1.86	-0.45	0.70	3.74	0.54	2.14
	Ex Tax	5.04	1.57	3.31	1.76	-0.47	0.64	3.52	0.44	1.98
High Discount Rate, High Travel	Total	2.34	-0.18	1.08	1.02	-0.86	0.08	2.42	-0.11	1.15
	Ex Tax	2.32	-0.08	1.12	0.96	-0.85	0.05	2.26	-0.18	1.04
High Discount Rate, High Travel, Low Sulphur Diesel (+2 cents/litre)	Total	2.53	0.00	1.27	1.02	-0.86	0.08	2.42	-0.11	1.15
	Ex Tax	2.51	0.11	1.31	0.96	-0.85	0.05	2.26	-0.18	1.04
Improved Fuel Economy 10% Lower Fuel Consumption	Total	2.73	0.79	1.76	0.45	-1.05	-0.30	1.49	-0.49	0.50
	Ex Tax	2.70	0.84	1.77	0.43	-1.04	-0.30	1.40	-0.54	0.43
Tax Credit for 10% of CNG Station Cost	Total	2.73	0.79	1.76	-0.04	-1.19	-0.62	0.98	-0.64	0.17
	Ex Tax	2.70	0.84	1.77	0.43	-1.04	-0.30	1.40	-0.54	0.43

Table A6.2
Cost of Fuel Switching in France
Sensitivity Analysis Results
Base case: Travel 13 800 km/year; 5% Discount Rate; All other data as in **Table 7.3**

Cost of Fuel Switching in US cents per km

Case		Diesel			CNG 2/3 Range			CNG Full Range		
		High	Low	Midpoint	High	Low	Midpoint	High	Low	Midpoint
Base	Total	0.21	-3.84	-1.82	0.60	-0.54	0.03	1.67	0.10	0.88
	Ex Tax	3.05	-0.54	1.25	0.65	-0.39	0.13	1.43	0.02	0.73
High Annual Travel 20 000 km/year	Total	-2.79	-6.11	-4.45	0.30	-0.71	-0.20	1.24	-0.14	0.55
	Ex Tax	0.62	-2.38	-0.88	0.38	-0.55	-0.08	1.05	-0.20	0.43
Low Annual Travel 10 000 km/year	Total	1.30	-4.02	-1.36	0.99	-0.40	0.30	2.33	0.38	1.35
	Ex Tax	3.96	-0.72	1.62	0.96	-0.28	0.34	1.98	0.24	1.11
High Discount Rate 30%	Total	3.37	-3.56	-0.10	1.88	0.07	0.98	3.70	1.10	2.40
	Ex Tax	5.64	-0.26	2.69	1.70	0.11	0.90	3.11	0.85	1.98
High Discount Rate, High Travel	Total	0.04	-5.14	-2.55	1.18	-0.27	0.46	2.61	0.57	1.59
	Ex Tax	2.91	-1.56	0.68	1.11	-0.18	0.47	2.20	0.40	1.30
High Discount Rate, High Travel, Low Sulphur Diesel (+2 US cents/litre)	Total	0.16	-5.01	-2.43	1.18	-0.27	0.46	2.61	0.57	1.59
	Ex Tax	3.03	-1.43	0.80	1.11	-0.18	0.47	2.20	0.40	1.30
Improved Fuel Economy 10% Lower Fuel Consumption	Total	0.59	-3.45	-1.43	0.65	-0.44	0.11	1.70	0.18	0.94
	Ex Tax	3.10	-0.48	1.31	0.68	-0.32	0.18	1.46	0.09	0.77
Diesel Tax at Gasoline Rate	Total	2.44	-1.61	0.42	0.60	-0.54	0.03	1.67	0.10	0.88
	Ex Tax	3.05	-0.54	1.25	0.65	-0.39	0.13	1.43	0.02	0.73

ANNEX 7

THE MARKET FOR AUTOMOTIVE DIESEL FUEL

This annex considers the market for diesel cars in Europe and the United States. Diesel is a relatively familiar alternative fuel, and experience with its introduction in passenger cars may help guide strategies to introduce other fuels.

Diesel engines differ from gasoline engines in several respects:

Advantages	**Disadvantages**
Longer life	Higher cost (per rated kW)
Higher energy efficiency	Larger, heavier (per rated kW)
Lower fuel cost	Potential cold weather problems
Less component failure	Higher frequency and cost of routine maintenance
Lower emissions of carbon	More noise
monoxide, VOCs and NO_x[1]	Higher particulate emissions

Many of the disadvantages would have been found in early light-duty gasoline vehicles as well; they are being solved for light-duty diesel as the technology matures. With more advanced technology, diesel engines will probably be able to match gasoline engine performance. Light-duty diesel engines are likely to remain more expensive than gasoline engines for a given performance level, at least until 2005.

Diesel fuel is consistently cheaper than gasoline on the world market. **Figure A7.1** shows spot market prices for the two fuels from 1975 to 1991. A litre of diesel fuel is usually cheaper than a litre of gasoline at the pump. **Figure A7.2** shows average prices in the OECD from 1978 to 1990. Despite the price gap, however, diesel cars have achieved minimal market shares in most countries.

Interest in the diesel engine grew following the 1973 and 1979 oil price rises. There are several probable reasons:

- For the driver, diesel fuel was generally cheaper than gasoline, and diesel cars consumed less fuel than gasoline cars.

---

1. Lower NO_x than gasoline cars not equipped with cataytic converters.

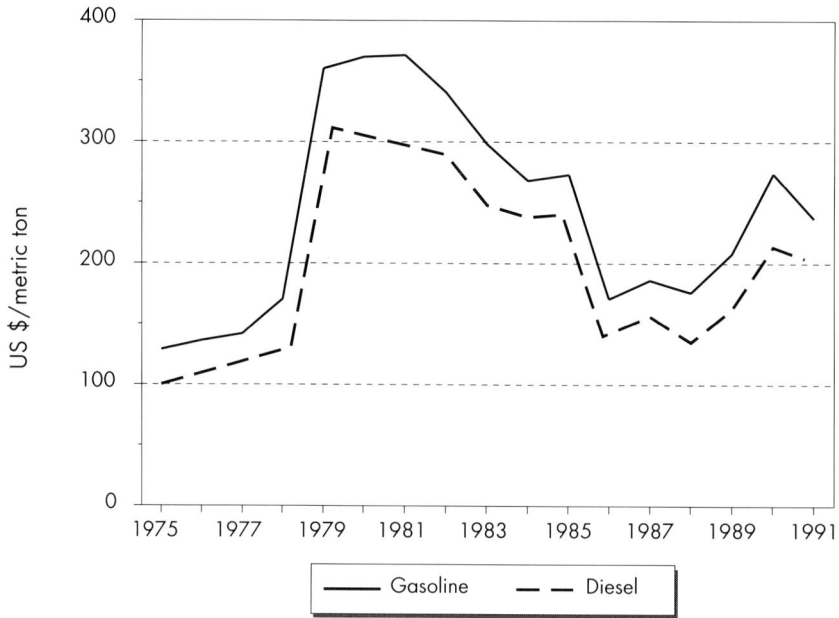

Figure A7.1
Rotterdam Spot Prices of Automotive Fuels, 1975 to 1991
(US$/metric ton)

Source: IEA Statistics.

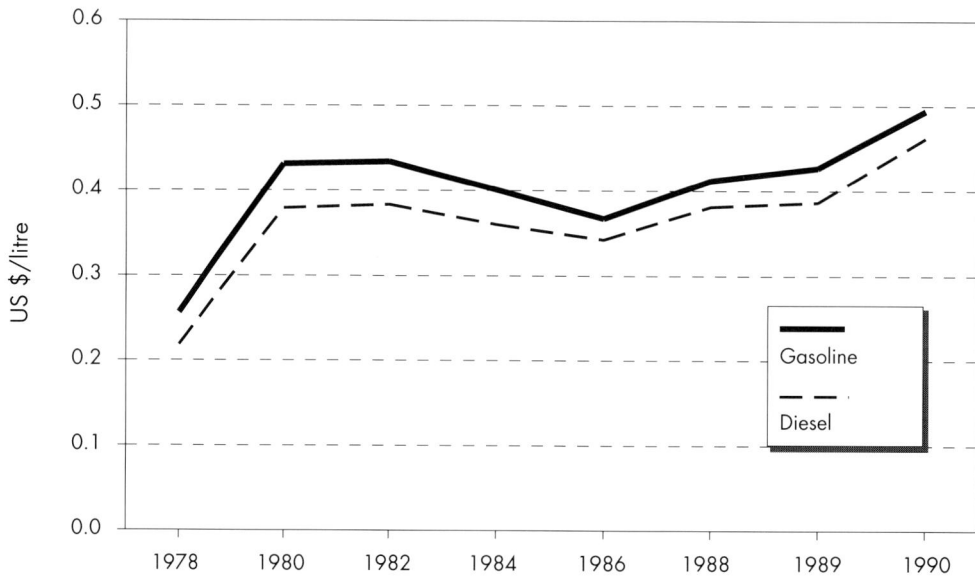

Figure A7.2
Average Retail Prices for Automotive Fuels, OECD, 1978 to 1990
(US$/litre)

Source: IEA Statistics.

- For car manufacturers, the diesel car represented a new market segment that could remain within a well-established fuelling network. Some additional profit could be earned by early entrants to the market.

- In the early 1980s, oil companies faced declining residential and industrial demand for gas oil, which is essentially the same product as diesel. A new market was needed for the fuel.

- Some governments saw the diesel car as a means of improving oil security and trade balances.

Light-duty diesel engines had been made in Europe since the 1930s. They were used mainly for commercial vehicles such as taxis and vans, where their durability and low fuel consumption in urban driving offered a considerable advantage. From about 1977, a number of US, Japanese and non-OECD manufacturers started to produce diesel cars. In Europe most car models are now available in a diesel version.

US Experience

In the United States, diesel was relatively cheap in the 1960s and 1970s, with a tax rate roughly half that on gasoline. During the 1970s manufacturers started to produce diesel cars to satisfy the Federal Government's CAFE requirement for an improvement in average new car fuel efficiency (Kurani and Sperling, 1988).

Diesel cars began to enter the market in 1977. By 1981 they made up over 6% of new car registrations. But by 1985 the market share had declined to almost zero. **Figure A7.3** shows the diesel-gasoline price difference and new diesel car sales from 1970 to 1991. The drop is generally

Figure A7.3
Diesel-Gasoline Price Difference and Sales of Diesel Cars, United States, 1970 to 1991

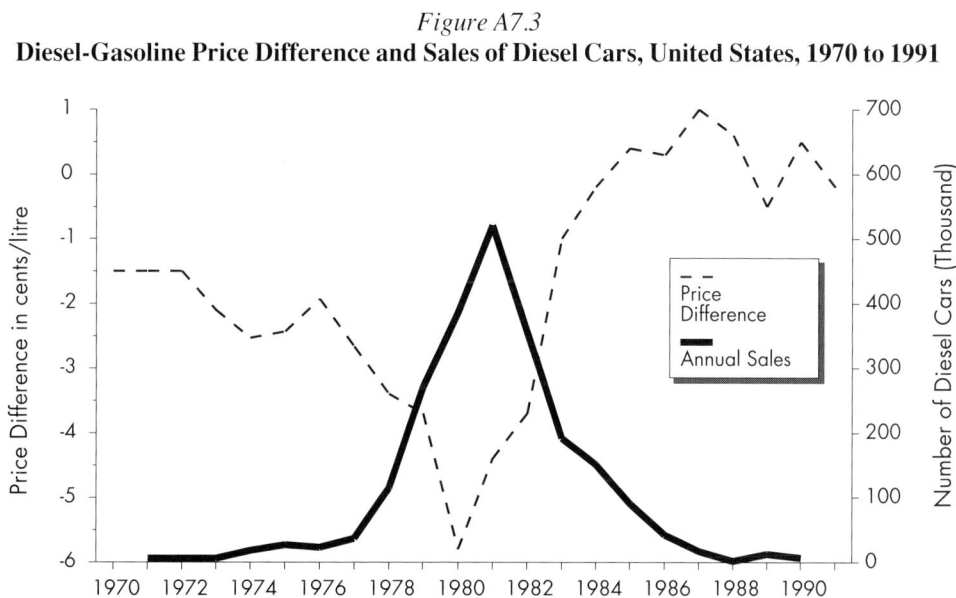

Source: Prices: IEA Statistics; Diesel Car Sales: Kurani and Sperling (1988).

ascribed to bad experience with one General Motors car model that was built with a truck diesel engine, which gave very poor driving characteristics. In addition, US oil distributors made relatively little attempt to attract customers to diesel, and car drivers generally had to use pumps intended for trucks — usually separate from the gasoline pumps, and often dirty.

During the early 1980s, US consumers perceived diesel cars to be cheaper to operate than gasoline cars. Then the fuel price advantage shrank after the tax on diesel was raised, which was probably another significant factor in the decline in market share (Kurani and Sperling, 1988). The difference in the cost per distance driven seems to have affected consumer choice less than the difference in the pump price. As the price gap narrowed, the share of diesel cars in the new car market declined. Diesel penetration of new car sales in France also slowed during periods of smaller gasoline/diesel price differences (DRI European Energy Services, 1989).

Experience in Europe

The diesel car market in Europe has developed much farther than that in the United States. In most of the European Community, diesel cars made up more than 10% of new registrations in 1988. In some countries the share exceeded 20%. In many European countries diesel vehicles make up the majority of the commercial light-duty fleet.

Diesel car prices are higher than those of gasoline cars with the same engine rating. The difference is only partly accounted for by higher costs and taxes, giving some indication of the buoyancy of demand for diesel cars. Demand in Europe has been stimulated by a number of factors:

- Diesel fuel has been taxed less than gasoline in most countries, both per litre and on an energy basis. In many countries this is intended to avoid disabling freight transport, but governments also see some value in the introduction of diesel cars.

- Manufacturers, including subsidiaries of US car makers in Europe, have seen diesel vehicles as a marketing opportunity and have worked hard to improve vehicle performance.

- Oil companies have supported a move to diesel by placing diesel pumps alongside gasoline pumps in filling stations and competing for market share by offering products such as deodorised diesel fuel.

Promotion of diesel vehicles has been strongest in France, where the price of diesel fuel typically was about 40% below that of gasoline during the 1980s. The purchase tax on diesel cars is lower than that on gasoline cars. The strong government support for diesel is due in part to the importance of the French automotive industry. French car manufacturers have become strong exporters of diesel cars and also supply engines to other manufacturers. In addition the approach reflects a more general concern in French energy policy to minimise oil use. The resulting uptake of diesel cars in France is shown in **Figure A7.4**.

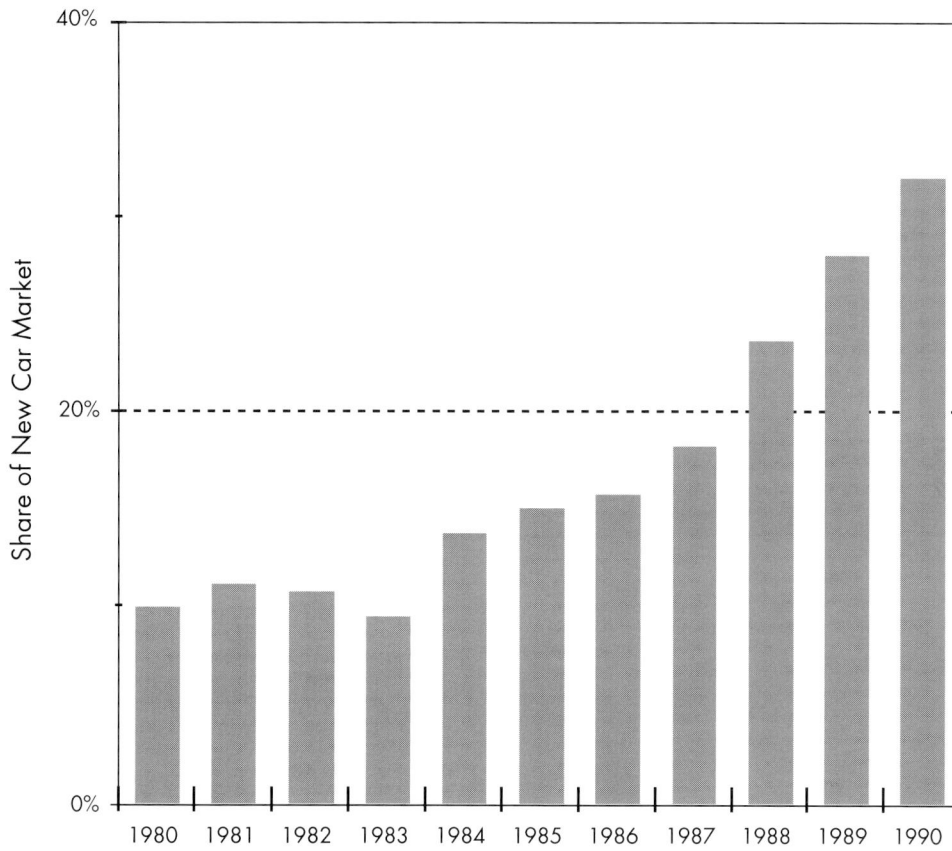

Figure A7.4
Diesel Share of New Cars Sales, France, 1980 to 1990
(%)

Source: DRI European Energy Services (1989).

ANNEX 8

THE MARKET FOR UNLEADED GASOLINE

Lead compounds — most commonly tetra-ethyl lead — have been added to gasoline as octane enhancers in OECD Member countries since the 1920s. Unleaded gasoline has been phased in since the 1970s in many OECD countries. In almost all cases its introduction has coincided with legislation requiring the use of catalytic converters.

United States

Unleaded gasoline was introduced in the United States as a result of the 1970 Clean Air Act, which effectively required catalytic converters to be used on gasoline cars.

While the Act specified ambient air quality standards for lead, no schedule for reduction of lead emissions was established (Sperling and Dill, 1987). General availability of unleaded gasoline was mandated by the EPA in 1973 to ensure that adequate supplies would exist by the time catalytic converters were introduced in 1975. Gasoline retail outlets selling more than 200 000 gallons a year were required to sell unleaded gasoline from 1st July 1974 (Sperling and Dill, 1987).

Automotive Industry Response. US car manufacturers, viewing catalytic converters as the best option for meeting the 1975 emissions standards, supported the move to unleaded gasoline. Despite delays in the introduction of the standards, car makers smoothed the transition to unleaded gasoline by making new cars compatible in the early 1970s. This meant reducing compression ratios, as unleaded gasoline at the time had a low octane rating. It also required hardening of valve seats, which had depended on lead as a lubricant. Roughly 40% of the fleet was compatible with unleaded gasoline by 1975, assuring a market for the oil companies.

Oil Industry Response. The oil industry faced major changes with the introduction of unleaded gasoline. Without lead additives, gasoline octane ratings could be maintained only by investing in more conversion plant in refineries or using additives such as MTBE. A complex system of fuel price regulations was established in the 1970s to help buffer refiner profits during the change (Sperling and Dill, 1987).

The overall extra cost to the consumer of supplying unleaded gasoline was estimated to be between $0.037 and $0.056 per gallon, mainly from investment in refineries, pipelines, distribution and storage facilities. The price difference between unleaded and leaded gasoline grew to exceed this as falling demand for leaded gasoline pushed prices down. On the wholesale market, unleaded gasoline cost $0.025 a gallon more than leaded in 1976 and $0.054 more in 1980; the difference stabilised at $0.085 in 1983 (Sperling and Dill, 1987, p. 50).

Both the price difference and vehicle performance were unfavourable for unleaded gasoline. No fiscal incentive was offered to the consumer to offset the extra cost or encourage a shift to unleaded gasoline. A significant level of misfuelling resulted.

Retailers were prohibited from providing leaded gasoline to drivers of vehicles requiring unleaded, and fined if they did so. At the same time, car manufacturers had to narrow the fuel tank inlets on cars requiring unleaded, and gasoline retailers had to install narrower nozzles on unleaded pumps. Manufacturers were also required to label cars. But misfuelling continued to occur frequently.

As a result of the 1990 Clean Air Act Amendments, sales of leaded gasoline will be banned from 31st December 1994. In 1990, 9.5% of the US car fleet was more than 15 years old (Davis and Morris, 1992). Some of these vehicles cannot use unleaded gasoline; any remaining at the beginning of 1995 are likely to be scrapped.

Europe

The approach to the introduction of catalytic converters and unleaded gasoline varies between countries in OECD Europe. Those in the Stockholm Group[1] had introduced catalytic converters by 1990. Unleaded gasoline makes up at least half of the car fuel market in these countries.

In the European Community, Directive 85/210/EEC required unleaded gasoline to be generally available by October 1989, in preparation for the introduction of catalytic converters. Member states are required to take steps to avoid misfuelling, and to adopt incentives to accelerate market penetration of unleaded gasoline. EC Directive 87/416/EEC allows Member states to ban sales of regular leaded gasoline, an option so far taken up only by Germany, Luxembourg and Belgium (CONCAWE, 1991).

To make unleaded gasoline cheaper at the pump than leaded, governments have provided tax differentials, ranging from US$0.02 to $0.134 a litre, between 95 octane unleaded gasoline and premium leaded. Retail price differences vary from $0.065 to $0.125.

Market shares for unleaded gasoline appear to be more affected by availability than by the size of the price difference. In most European markets the uptake of unleaded has been very rapid. Little misfuelling is known to have occurred.

---

1.   Austria, Demark, Finland, Norway, Sweden and Switzerland.

ANNEX 9

TRAFFIC MODELLING IN THE NETHERLANDS[1]

This case study examines system-wide changes in the transport sector for the Randstad, a region of the Netherlands that includes several cities. Policies and policy packages are assessed, as means of influencing traffic growth and energy consumption, to see their effects on CO_2 emissions. The packages considered are designed to affect traffic patterns and travellers' choices of transport modes and routes. Such measures would interact with other elements of government policy. The measures might be designed mainly to improve safety or urban air quality or to reduce traffic congestion. This study does not consider the effectiveness of the measures in achieving these objectives. A transport system model is used to estimate the effects of measures on energy consumption and hence on CO_2 emissions.

Background — Previous Studies

The Randstad region in the western Netherlands includes Rotterdam, Utrecht, Amsterdam and The Hague. Traffic congestion in the region is growing. Traffic flows in the Netherlands are projected to increase by 70% between 1990 and 2010. The Government is seeking policy strategies to reduce the growth to about 35%. Several studies of the transport system and policies' probable effects have been undertaken. The results of some of these studies have been used for an analysis of possible strategies for the Randstad.

One study for The Hague, commissioned by Projectbureau Energy Research (now NOVEM) in 1988, showed that:

- parking restrictions in the city centre could reduce energy use both in aggregate and per vehicle-kilometre;

- traffic restrictions in residential areas would result in an overall increase in traffic and congestion because drivers would lengthen journeys to bypass restricted zones;

- traffic signalling improvements, such as urban traffic control or area-wide traffic control, improve traffic flow but attract additional drivers, resulting in an overall increase in energy use.

---

1. Based on a report prepared for the IEA by NOVEM (1992).

A study commissioned in 1988 by the Ministry of Transport examined the effects of improved public transport on traffic on the A4 highway near the Amsterdam airport. At the time, 78.4% of passenger traffic was by car and 21.6% by public transport. The study showed that:

- improvements in public transport could, in principle, change the split to 74.2% public transport and 25.8% private car, though the level of public transport service required would be unrealistic;

- achievable levels of improvement to public transport could increase its share to 52.7%.

A survey on passenger attitudes affecting the choice of public transport in The Hague in 1989 indicated that:

- although the relative costs of public transport and car use affect modal choice, travel times tend to have more influence;

- passengers rate each transfer on public transport as equivalent to 7.5 minutes of extra travel time;

- travel time by car has only about half the significance in modal choice as travel time by public transport;

- the time cost of transferring from car to public transport at a park-and-ride facility is perceived as equivalent to 15 minutes in a car or 7.5 minutes on public transport.

Methodology for the Randstad Study

A transport network model was used to examine the impact of policy measures on traffic and hence on CO_2 emissions in the Netherlands. The Netherlands and Belgium are represented in the model by 958 zones, each treated as a starting point and destination for trips. The Randstad region is broken down into 724 very small zones and Belgium into ten large zones.

The model uses trip generation and attractor factors based on population density, car availability and employment patterns. Different factors are used for metropolitan areas, small towns and rural areas. Separate factors are used for trips that are work-related and those that are not. The factors are used to calculate the number of trips starting from, and ending at, each zone during the weekday afternoon peak hour.

Within the model, the trips are assigned between modes on the basis of trip "resistance". More travellers will use lower resistance modes. The resistance of each mode for a given trip is determined by the financial cost and time required. Travellers can use cars, bicycles or public transport. Transfer times, parking fees and time spent looking for parking spaces are included in the resistance for each journey.

Travel times and costs between each pair of zones are determined by a model of the road and public transport networks. Journeys can be assigned to routes in the networks by various means, taking account of route length, capacity and delays.

The model calculates, as its main output, the number of journeys between each pair of zones, by each mode, and according to trip purpose. The output has been calibrated against actual observations of travel behaviour to obtain corrections to the trip generation and attraction factors.

CO_2 emissions from the network can be calculated from traffic flows, using assumed fuel consumption rates. In this study, it is assumed that cars consume 6 litres/100 km in steady traffic flows and 1.7 litres/hour during congestion delays. The average consumption is 7.19 litres/100 km.

Policy Measures and Packages

The fully configured model can be used to test the effects of transport policy measures, such as increases in costs for a given mode or constraints on road use or parking. The 14 policy measures considered in the study are shown in **Table A9.1**. All the measures are currently used to some degree in Dutch transport programmes. The individual measures have been grouped into six packages, each of which would constitute a coherent policy strategy:

- "Parking Control" package: reducing parking space and raising fees, at the same time improving public transport;

- "Price Policy" package: imposing road pricing and fuel tax while also improving public transport;

- "Do-Nothing" package: not adding road infrastructure but improving public transport to meet demand;

- "Meet-Demand" package: improving road capacity to meet demand while also improving public transport;

- "Safety" package: restricting car use and speed in residential areas and cities, and improving public transport and bicycle facilities (not modelled);

- "Combination" package: providing park-and-ride facilities and improving public transport (not modelled).

The study incorporates government policies designed to influence transport development. It is estimated that, to 2010, these policies would require a budget of 13.6 billion guiders (Gld)[1]. Half of this amount is assumed in the reference case to be used for maintenance and basic improvements. The increase in public transport specified for the model is estimated at Gld 7 billion, with the remainder for maintaining and improving the road network. In 1987, taxes represented 69.3% of the final price of gasoline, 48.1% for diesel fuel. The price of fuel in the model is Gld 1.63/litre. Parking measures are budget neutral, the extra revenue financing the additional enforcement needed.

Modelling Results

Table A9.2 presents the policy packages, their costs and their expected effects on CO_2 emissions from cars. Changes in emissions from public transport are not included. The packages are not generally designed to reduce CO_2 emissions so the costs should not be ascribed to CO_2 abatement.

---

1.   On average in 1992, Gld 1 = $0.569.

Table A9.1
Effects of Measures in Transport Model for the Netherlands, 2010

Case	Description	Traffic Index	Speed Index	CO_2 Index
0	Reference Case	100.0	100.0	100.0
1	Double parking fees, extend parking fee area	99.1	101.5	99.0
2	Limit parking accomodation	98.7	100.6	98.8
3	Increase fuel price by 30%	93.9	99.1	94.3
4	Increase road pricing (raise driving cost by 50% in cities, 25% elsewhere)	92.6	99.1	93.3
5	Charge Gld 5 tolls to enter city area	99.1	98.3	99.4
6	Do not build new roads to meet increasing traffic demand	98.6	103.4	98.5
7	Provide roads to meet demand	108.3	98.5	107.3
8	Limit traffic in residential areas		not modelled	
9	Improve public transport	99.4	100.2	99.5
10	Provide park-and-ride facilities		not modelled	
11	Designate lanes for high-occupancy cars	96.8	100.6	96.4
12	Improve traffic management		not modelled	
13	Lower speed limit by 30%	88.1	85.4	96.0
14	Lower speed limit by 10%	98.4	95.7	100.3

Table A9.2
Effects of Policy Packages in Transport Model for the Netherlands, 2010

Case	Description	Traffic Index	Speed Index	CO_2 Index	Cost (billion Gld)
1	Parking Control Package (higher fees, limited area plus improved public transport)	94.3	102.4	94.4	7
2	Price Policy Package (fuel price increase, road pricing and parking fees plus improved public transport)	83.7	102.1	83.3	7
3	Do-Nothing Package (no investment in roads but improve public transport)	94.6	103.9	94.7	7
4	Meet-Demand Package (invest in roads to meet demand and improve public transport)	104.6	97.4	104.3	14

Parking control measures can have unexpected results and have to be designed with care. People making short trips are more likely to be discouraged by parking difficulties than those making long trips. Displacing short-trip traffic, where parking capacity is a constraint, creates more parking space for long-trip traffic. The overall effect of the parking control package is to reduce CO_2 emissions by about 6%.

The price-policy package is the most effective one, reducing emissions by almost 17%. This package, however, would require the introduction of a new pricing system, which could prove to be politically difficult. Fuel price increases result in a greater decrease in vehicle-kilometres overall than in the urban area. Higher fuel and road use costs tend to reduce long trips, freeing road capacity in the urban centres for more short trips. Improved public transport mainly attracts long-distance commuters and so affects the system as a whole more than urban areas.

The "do-nothing" package reduces CO_2 emissions by 15% in urban areas, as increased congestion discourages car use, but only 5% overall. Emissions increase by 4% in the "meet demand" package, where new road infrastructure is provided, since this encourages car use.

ANNEX 10

GREENHOUSE GAS EMISSIONS FROM AUSTRALIAN TRANSPORT[1]

Aim of Study

In October 1990 the Federal Government adopted an interim planning target for greenhouse gas emissions, with the stipulation that, in the absence of similar action by other major greenhouse gas producing countries, stabilisation should not hurt the economy. The target is stabilisation at 1988 levels by 2000 and a 20% reduction by 2005. This case study focuses on the possibility of achieving reductions of this order in the transport sector through energy efficiency improvements, the use of alternative fuels, transport system improvements, better land use and reductions in discretionary travel. The study discusses the problems of estimating the costs and benefits of a greenhouse gas abatement strategy that includes these approaches.

Energy Consumption and Greenhouse Gas Emissions

The estimated share of transport in total primary energy consumption in Australia in 1987/88 was 27.3%. Transport's share in final energy consumption in that fiscal year was 37% and its contribution to total Australian greenhouse gas emissions was estimated at 14%. Other sectors produce considerable amounts of non-CO_2 greenhouse gases, such as methane and N_2O. The main component of anthropogenic greenhouse gas emissions is CO_2, of which domestic transport contributed roughly 26% in 1987/88. The car accounts for about 54% of transport sector CO_2 emissions, and 39% can be attributed to urban use of private cars. **Figure A10.1** shows CO_2 emissions from domestic transport by vehicle type for 1987/88. Urban passenger transport accounts for 45% of emissions, non-urban passenger transport 24%, non-urban freight 19% and urban freight 12%.

The expected rate of increase in fuel demand in the 1990s is 2-2.2% a year. This is lower than past growth levels: from 1976 to 1988 the annual growth rate was about 4%. Emission scenarios based on fuel demand growth rates of 1%, 2% and 3% show that, to attain an absolute emission reduction of 20% from 1988 levels in a business-as-usual scenario, actual emission reductions of

---

1. Based on material provided to the IEA by the Australian Government.

CO_2 Emissions from Australian Domestic Transport, 1987/88
(%)

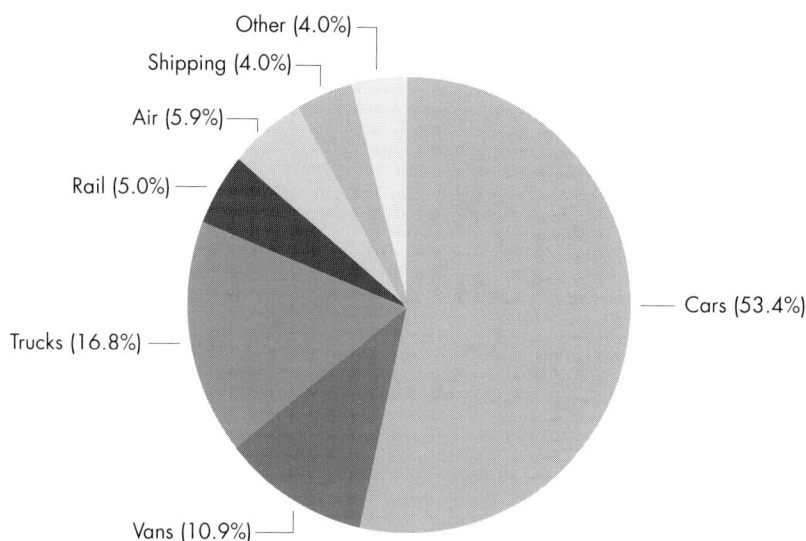

Other (4.0%)

Shipping (4.0%)

Air (5.9%)

Rail (5.0%)

Trucks (16.8%)

Cars (53.4%)

Vans (10.9%)

Note: "Other" consists mainly of buses, motorcycles and light aircraft.

between 30% and 50% from the projected emission levels for 2005 would be required. An underlying trend in technical change is built into these forecasts, implying fleet fuel economy improvements of as much as 15% from 1988 to 2005. This gives an indication of the magnitude of the challenge involved in reducing greenhouse gas emissions in the Australian transport sector.

Assessment of Emission Reduction Measures

The study notes that, as the marginal greenhouse gas abatement cost varies between different parts of the transport sector, the most cost-effective way to reduce emissions from transport would be to cut emissions in those areas where the abatement cost is the lowest.

The Government recommends that any strategy adopted should be cost-effective and preferably justifiable on broad grounds of efficiency, or offer economic or environmental benefits besides greenhouse gas reductions. As the study emphasises, however, there are difficulties involved in assessing the costs and benefits to the economy of greenhouse gas reduction strategies. Some comment is included on the effects of strategies in particular industries.

Potential Fuel Economy Measures. As cars contribute 54% of Australia's transport CO_2 emissions, a key component in the national strategy to reduce greenhouse gas emissions must be to improve the fuel efficiency of cars. There are diverging views, however, on the technological potential to do so. The Australian motor industry has proposed fuel economy targets of 8.2 litres/100 km by 2000 and 8 litres/100 km by 2005, compared to the 1988 fleet average of 11.8 litres/100 km. The target for 2000 roughly corresponds to the 1988 US average for new cars; improvements in fuel economy may be more costly for the Australian automotive industry because it has fewer economies of scale in production.

Diesel engines are expected to power merchant ships for the foreseeable future. The most substantial fuel economy improvement could come from better fuel injection systems in these engines. Ship fuel efficiency can be further improved by better hull design. By 2005, the fuel consumption rate per 1 000 metric tons of dead weight in the Australian flag fleet could improve by 15-25%.

Significant aircraft energy efficiency improvements will arise from fleet turnover, with continuing improvements in new aircraft efficiency in the period to 2005. A 10-15% improvement in the Australian aircraft fleet's average fuel efficiency could be expected by 2005, compared to 1988.

Alternative Fuels. Diesel vehicle use is likely to be constrained by controls on particulate and NO_x emissions. Availability of diesel from Australian refineries may be limited. Use of LPG is not expected to exceed 5% of transport fuel, also because of limited supply. The initial use of methanol may be to produce MTBE as an octane enhancer for gasoline. By 2005 CNG may substitute for some 10% of diesel fuel in road transport. Its main use now is for buses and trucks. If the CNG distribution system were improved and excise taxes were not unfavourable, greater penetration would be possible, especially for long-distance transport.

The use of ethanol from sugar cane and grain will be limited since supplies are short and the economics using current technology are poor. Vehicles consuming hydrogen produced from solar-generated electricity would need at least another 20 years of development before commercialisation. As for electric vehicles, if battery life could be prolonged, rapid recharging developed, performance improved and costs brought down, households might use electric "second cars" by 2000.

Transport System Improvements

Passenger Transport. Cars accounted for around 88% of CO_2 emissions from urban passenger transport in 1987/88. Commuting for work produces the highest rates of emissions from cars because of traffic congestion and low occupancy. Such trips account for 35% of urban car traffic. Many drivers do not have a convenient alternative, especially since only about 6% of urban passenger kilometres are journeys to work in central areas. Car occupancy rates are higher and congestion less for other uses of cars. The energy intensity of off-peak car use may not differ greatly from that of public transport, which has low off-peak occupancy levels.

Policy measures might therefore focus on decreasing emissions during peak traffic periods where travel to and from work dominates. This might be done through traffic management, parking measures, price measures, improved commuter facilities and expanded public transport.

In non-urban areas the car was responsible for roughly 63% of transport CO_2 emissions in 1987/88. Energy consumption could be reduced by lowering speed limits, but this is politically difficult. Modal substitution may also be difficult in non-urban areas. The study finds that the most energy-efficient mode is the bus, which achieves higher load factors and faces better traffic conditions than in urban areas. The most energy-intensive mode is air transport, which is not likely to be replaced by other modes for the trip lengths involved. A major source of emission reduction will be technological developments in fuel economy.

Freight Transport. Non-urban freight accounts for around 19% of total transport emissions. Long-haul bulk freight is mainly transported by sea or rail. Given the complex nature of urban freight transport, it is unlikely that this task could be efficiently performed with a very different vehicle mix. The majority of non-bulk freight is carried by road. Some switching to rail would be possible, but major structural improvements would be required for rail to offer the flexibility of road transport. Road pricing in line with economic and social costs would improve the economics of rail freight. Fuel economy improvements may also be available within each mode, through technological change and higher load factors.

Improving Land-Use Planning

In terms of population density, the major Australian cities resemble those of North America. Low residential density may be the most important element underlying extensive use of private cars and high urban gasoline consumption. Growth in car use might be reduced through measures that improve land-use planning by increasing residential density and concentrating urban development on "corridors" that are easily served by public transport.

Evaluation of Strategies

The study concludes that it is not clear to what extent improvements in car technology will reduce overall fuel consumption. Changes in driving conditions and driver behaviour might increase actual energy use. It is hard to assess the cost of policy measures to improve fuel efficiency, since factors such as consumer benefit, safety and social costs must be taken into account. If opportunities to change vehicle technology become limited by the relatively short time available to reach greenhouse gas reduction targets, car use might have to be reduced. This could be done through measures that also reduce urban congestion.

Alternative fuels may offer lower emissions, but they are considered to have limited market potential without some significant technological breakthrough.

The scope for changes in the transport system to lead to modal shifts is unresolved. There is some potential for freight to shift to rail. Urban car use might be reduced in favour of public transport, perhaps primarily to reduce congestion and improve safety.

If the greenhouse gas emissions reduction target is attained through policies that offer parallel benefits and require little capital outlay, the cost for Australian industry will not be high. If policy measures are used that raise freight costs and travel costs in general, transport-intensive industry exposed to international competition could be adversely affected. Such effects might be mitigated by international reciprocity.

The effect on the Australian motor industry of measures to improve vehicle fuel economy depends on the measure used. Early scrappage may benefit new car sales, while measures that increase retooling costs would have an adverse effect. Much would depend on policies of overseas parent companies.

Conclusion

Given the difficulty in achieving government greenhouse gas abatement targets, the potential for abatement in the transport sector requires careful evaluation. There is considerable uncertainty about the cost of reducing emissions from cars, and further analysis is needed.

Some gradual greenhouse gas abatement can be achieved through energy efficiency improvements in vehicles, use of alternative fuels, transport system improvements (including modal shifts) and improved urban planning. Rapid emission abatement from the transport sector may, however, require reductions in travel.

REFERENCES

American Petroleum Institute [API] (1990). *The Economics of Alternative Fuel Use: Compressed Natural Gas As a Vehicle Fuel.* American Petroleum Institute Research Study #056, Washington DC.

ARCO (1991). Private Communication, August.

Beck, N.J., W.E. Waseloh, R.L. Barkhimer and W.P. Johnson (1984). *CRIDEC: A Practical Solution to High-Performance, All Electronic Fuel Injection for Diesel Engines.* Society of Automotive Engineers, Inc., Technical Paper Series, No. 845033, Warrendale, PA.

Bennethum, J.E. and R.E. Windsor (1991). *Toward Improved Diesel Fuel.* Society of Automotive Engineers, Inc., Technical Paper Series, No. 912325, Warrendale, PA.

British Rail (1993). Private Communication, January.

Button, K. (1990). "Environmental Externalities and Transport Policy", *Oxford Review of Economic Policy.* Vol. 6., No. 2.

California Air Resources Board [CARB] (1992). *Technical Feasibility of Reducing NO_x and Particulate Emissions from Heavy Duty Engines.* Mimeo of draft report.

California Energy Commission (1991). *Cost and Availability of Low-Emission Motor Vehicles and Fuels.* Draft, AB234 Report Update, August.

Cambridge Energy Research Associates [CERA] (1992). *The US Clean Air Act: Reshaping the Downstream.* Cambridge, MA.

Canadian Energy Research Institute [CERI] (1992). Private Communication, July.

Cannon, J.S. (1989). *Drive for Clean Air, Natural Gas and Methanol Vehicles.* Inform, Inc., New York, NY.

Clean Fuels Report (The) (1991). Vol. 3, No. 2, J.E. Sinor Consultants Inc.

227

Clean Fuels Report (The) (1992). Vol. 4, No. 5, J.E. Sinor Consultants Inc.

Commission of the European Communities [CEC] (1991). *Proposal for a Council Directive Relating to the Sulphur Content of Gasoil.* COM(91)154 final, SYN 340, Brussels, 10th June.

Commission of the European Communities [CEC] (1992). *Research and Technology Strategy to Help Overcome the Environmental Problems in Relation to Transport.* Directorate General for Science, Research and Development, SAST Project No. 3, Global Pollution Study, EUR-14713-EN, Brussels and Luxembourg.

CONCAWE (1991). *Motor Vehicle Emission Regulations and Fuel Specifications — 1990 Update.* CONCAWE Report No. 3/91, Brussels.

Davis, S.C. and M.D. Morris (1992). *Transportation Energy Data Book: Edition 12.* Prepared by Oak Ridge National Laboratory, Oak Ridge, TN, for US Department of Energy, Washington, DC.

DeLuchi, M.A. (1992). *Emissions of Greenhouse Gases from the Use of Transportation Fuels and Electricity.* Centre for Transportation Research, Argonne National Laboratories, Argonne, IL.

DeLuchi, M.A., R.A. Johnston and D. Sperling (1988). *Methanol vs. Natural Gas Vehicles: A Comparison of Resource Supply, Performance, Emissions, Fuel Storage, Safety, Costs, and Transitions.* Society of Automotive Engineers, Inc., Technical Paper Series, No. 881656, Warrendale, PA.

DeLuchi, M.A., Q. Wang and D. Sperling (1989). "Electric Vehicles: Performance, Life-Cycle Costs, Emissions and Recharging Requirements", *Transportation Research-A.* Volume 23A.

Department of Transport (1992). *Transport Statistics Great Britain 1992.* Government Statistical Service, London.

Difiglio, C., K.G. Duleep and D.L. Greene (1990). "Cost Effectiveness of Future Fuel Economy Improvements", *The Energy Journal.* Vol. 11, No. 1.

DRI European Energy Services (1989). *European Transport Energy Consumption.* DRI/McGraw-Hill, Lexington, MA.

Ecotraffic (1992). *The Life of Fuels.* Stockholm.

Energy and Environmental Analysis [EEA] (1991a). *An Assessment of Potential Passenger Car Fuel Economy Objectives for 2010.* Draft Final Report Prepared for US Environmental Protection Agency, Air and Energy Policy Division, Arlington, VA.

Energy and Environmental Analysis [EEA] (1991b). *Fuel Economy Technology Benefits.* Presented to the Technology Subgroup, Committee on Fuel Economy of Automobiles and Light Trucks, Detroit, MI, 31st July 1991.

Energy, Mines and Resources Canada (1988). *1988 National Survey of Auto-Propane Users.* Transportation Energy Report No. TE-88-8, Ottawa, Ontario.

Energy, Mines and Resources Canada (1990). *Natural Gas for Vehicles: Industry Survey 1989.* Transportation Energy Report No. TE 90-01, Ottawa, Ontario.

European Conference of Ministers of Transport [ECMT] (1990). *Transport Policy and the Environment.* OECD, Paris.

Europe Environment (1993). *Air Pollution: Tougher Car Emission Standards Proposed.* No. 401, January 7.

Faiz, A. (1991). *Automotive Emissions in Developing Countries — Relative Implications for Global Warming, Acidification and Urban Air Quality.* World Bank, Washington, DC.

Frankl, G., B.G. Barker and C.T. Timms (1988). *Electronic Unit Injectors.* Society of Automotive Engineers, Inc., Technical Paper Series, No. 885013, Warrendale, PA.

Gabel, H.L. and L.H. Röller (1992). "Trade Liberalization, Transportation and the Environment", *The Energy Journal.* Vol. 13, No. 3.

Gawthorpe, R.G. (1983). "Train Drag Reduction from Simple Design Changes", *International Journal of Vehicle Design.* Special Publications SP3.

Golob, T.B. et al. (1991). "Predicting the Market Penetration of Electric and Clean-Fuel Vehicles", Paper presented at INRETS International Symposium on Transport and Air Pollution, Avignon, France, 10th-13th September.

Greene, D.L. (1990). *Commercial Aircraft Fuel Efficiency Potential Through 2010.* Oak Ridge National Laboratory, Oak Ridge, TN.

Hadder, G.R., S. Das, R. Lee, N. Domingo and R.M. Davis (1991). "Ultra Low Sulphur Diesel Fuel Impacts on Availability of Other Fuels", *Energy Policy.* June.

Heath, M. (1991). *Alternative Transportation Fuels.* Canadian Energy Research Institute, Study No. 37, Calgary, Alberta.

Held, W., A. Konig, T. Richter and L. Puppe (1990). *Catalytic NO_x Reduction in Net Oxidizing Exhaust Gas.* Society of Automotive Engineers, Inc., Technical Paper Series, No. 900496, Warrendale, PA.

Hough, A.M. and C.E. Johnson (1990). *Modelling the Role of Nitrogen Oxides, Hydrocarbons and Carbon Monoxide in the Global Formation of Tropospheric Oxidants.* Harwell Laboratory, Report R 13545, Harwell.

IEA Coal Research (1984). *The Economics of Producing Ammonia and Hydrogen.* An Economic Assessment Report, London.

Innas (1992). *Energy Consumption and Greenhouse Gases from Heavy Duty Road Vehicles.* Consultants Report to the IEA, Breda.

Institute of Applied Energy (1991). *Current Status of Electric Vehicle Research and Development and Possibility of International Collaboration.* Interim report prepared for IEA, Tokyo.

Intergovernmental Panel on Climate Change [IPCC] (1990). Climate Change: The IPCC Scientific Assessment. Cambridge University Press.

Intergovernmental Panel on Climate Change [IPCC] (1992). Climate Change 1992: The Supplementary Report to The IPCC Scientific Assessment. Cambridge University Press.

International Energy Agency [IEA] (1990a). *Energy and the Environment: Policy Overview.* OECD, Paris.

International Energy Agency [IEA] (1990b). *Substitute Fuels for Road Transport.* OECD, Paris.

International Energy Agency [IEA] (1991a). *Energy Efficiency and the Environment.* OECD, Paris.

International Energy Agency [IEA] (1991b). *Energy Policies of IEA Countries: 1990 Review.* OECD, Paris.

International Energy Agency [IEA] (1991c). *Fuel Efficiency of Passenger Cars.* OECD, Paris.

International Energy Agency [IEA] (1991d). *Greenhouse Gas Emissions: The Energy Dimension.* OECD, Paris.

International Energy Agency [IEA] (1992a). *Energy Balances of OECD Countries 1989-1990.* OECD, Paris.

International Energy Agency [IEA] (1992b). *Energy Statistics of OECD Countries 1989-1990.* OECD, Paris.

International Energy Agency [IEA] (1993). *Energy Technology Strategy 21.* OECD, Paris (forthcoming).

International Road Federation [IRF] (1990). *World Road Statistics 1985-1989.* Geneva and Washington, DC.

Johnson, C.E. and J. Henshaw (1991). *The Impact of Emissions from Tropospheric Aircraft.* Harwell Laboratory Report AEA-EE-0127, Harwell.

Johnson, C.E., J. Henshaw and G. McInnes (1992). "The Impact of Aircraft and Surface Emissions of Nitrogen Oxides on Troposheric Ozone and Global Warming", Nature. Vol. 355, 2nd January.

Kroon, M. (1992). "Vehicle Selfcontrol; Engine Downsizing as a Prerequisite for Sustainable and Safe Traffic", Paper presented at ECMT International Seminar on Reducing Transport's Contribution to Global Warming, Paris, September 30th.

Kurani, K.S. and D. Sperling (1988). "Rise and Fall of Diesel Cars: A Consumer Choice Analysis", *Transportation Research Record.* Vol. 1175.

Lange, W.W. (1991). *The Effect of Fuel Properties on Particulates Emissions in Heavy-Duty Truck Engines under Transient Operating Conditions.* Society of Automotive Engineers, Inc., Technical Paper Series, No. 912425, Warrendale, PA.

Lloyds (1990). *Marine Exhaust Emissions Research Programme: Steady State Operation.* Lloyds' Register of Shipping, London.

McKeough, K. (1991). *Reformulated Gasoline.* Energy Development Information Development Note No. 28, Industry and Energy Department, Policy Research and External Affairs, World Bank, Washington, DC.

Marintek (1991). *Exhaust Gas Emissions from Ships in Norwegian Coastal Waters.* Consultants Report to the IEA, Norwegian Marine Technology Research Institute A/S, Trondheim.

Marrow, J.E., J. Coombs, and E.W. Lees (1987). *An Assessment of Bio-Ethanol as a Transport Fuel in the UK.* Vol. 1, Department of Energy, London.

Marrow, J.E. and J. Coombs (1990). *An Assessment of Bio-Ethanol as a Transport Fuel in the UK.* Vol. 2, Department of Energy, London.

Martin, D.J. and R.A.W. Shock (1989). *Energy Use and Energy Efficiency in UK Transport up to the Year 2010.* Department of Energy, London.

Miyaki, M., H. Fujisawa, A. Masuda and Y. Yamamoto (1991). *Development of New Electronically Controlled Fuel Injection System ECD-U2 for Diesel Engines.* Society of Automotive Engineers, Inc., Technical Paper Series, No. 910252, Warrendale, PA.

Moreno Jr, R. and D.G.F. Bailey (1989). *Alternative Transport Fuels from Natural Gas.* World Bank Technical Paper No. 98, IBRD, Washington, DC.

Needham, J.R., D.M. Doyle, S.A. Faulkner and H.D. Freeman (1989). *Technology for 1994.* Society of Automotive Engineers, Inc., Technical Paper Series, No. 891849, Warrendale, PA.

Newberry, D. (1990). "Pricing and Congestion: Economic Principles Relevant to Pricing Roads", *Oxford Review of Economic Policy.* Vol. 6, No. 2.

New Fuels Report (1991a). "Adopting the California Low Emission Vehicle Program in the Northeast States: An Evaluation", 15th July.

New Fuels Report (1991b). "Connecticut Becomes Third State To Adopt California Low-Emission Vehicle Plan", 20th May.

New Fuels Report (1991c). "Massachusetts, New York Seen as First to Opt into California Vehicle Program", 7th January.

New Fuels Report (1992). Vol. 13, No. 40.

Nishizawa, K., H. Ishiwata and S. Yamaguchi (1987). *A New Concept of Diesel Fuel Injection: Timing and Injection Rate Control System.* Society of Automotive Engineers, Inc., Technical Paper Series, No. 870434, Warrendale, PA.

NOVEM (1992). *Transport Policy, Traffic Management, Energy and Environment.* Consultants Report to the IEA, Netherlands Agency for Energy and the Environment, Utrecht.

Office of Technology Assessment [OTA] (1991). *Improving Automobile Fuel Economy: New Standards, New Approaches.* OTA-E-504, Washington, DC.

Okken, P. (1992). Private Communication, Netherlands Energy Research Foundation (ECN), August.

Organisation for Economic Co-operation and Development [OECD] (1991a). *Low Consumption/Low Emission Automobile.* Proceedings of an Expert Panel, Rome, 14th-15th February 1990.

Organisation for Economic Co-operation and Development [OECD] (1991b). *OECD Environmental Data: Compendium 1991.* Paris.

Organisation for Economic Co-operation and Development [OECD] (1993). *Choosing an Alternative Transport Fuel: Air Pollution and Greenhouse Gas Impacts.* Paris (forthcoming).

Petroleum Industry Research Associates [PIRA] (1990). Private Communication, July.

Porter, B.C. et al. (1991). *Engine and Catalyst Strategies for 1994.* Society of Automotive Engineers, Inc., Technical Paper Series, No. 910604, Warrendale, PA.

Ricardo (1992). *Study of Future Engine and Technology Requirements.* Consultants Report to the IEA, Ricardo Consulting Engineers Ltd., Shoreham-by-Sea.

Rigaud, G. (1989). "Influence du Prix de l'Energie Petrolière sur le Marché des Transports", Paper presented at 14th Congress of World Energy Conference, Montreal, Quebec, 17th-22nd September 1989.

Runzheimer (1993). Private Communication, January.

Schipper, L. (1992). Private Communication, November, Lawrence Berkeley Laboratory.

Schoubye, P., K. Pedersen, P.S. Pedersen, O. Grone and O. Fanoe (1987). *Reduction of NO_x Emissions From Large Diesel Engines.* CIMAC Paper D31.

Schumann, U. (ed.) (1990). *Air Traffic and the Environment — Background Tendencies and Potential Global Atmospheric Effects.* Springer-Verlag.

Sperling, D. and J. Dill (1987). "Unleaded Gasoline in the United States: A Successful Model of System Innovation", *Transportation Research Record*. Vol. 1175.

Stephenson, J. (1991). *Learning from Experiences with Compressed Natural Gas as Vehicle Fuel.* Centre for the Analysis and Dissemination of Demonstrated Energy Technologies, CADDET Analyses Series No. 5, Sittard.

Stumpp, G., W. Polach, N. Muller and J. Warga (1989). *Fuel Injection Equipment (FIE) for Heavy Duty Diesel Engines for US 1991/1994 Emission Legislation.* Society of Automotive Engineers, Inc., Technical Paper Series, No. 890851, Warrendale, PA.

Sypher:Mueller International (1992). *Methane as a Motor Fuel.* Report prepared for IEA, Ottawa, Ontario.

Tritthart P. and P. Zelenka (1990). *Vegetable Oils and Alcohols: Additive Fuels for Diesel Engines.* FISITA Paper 905112.

US Department of Energy [USDOE] (1988). *Assessment of Costs and Benefits of Flexible and Alternative Fuel Use in the US Transportation Sector, Progress Report Two: The International Experience.* Washington, DC.

US Department of Energy [USDOE] (1990a). *Assessment of Costs and Benefits of Flexible and Alternative Fuel Use in the US Transportation Sector, Technical Report Four: Vehicle and Fuel Distribution Requirements.* Washington, DC.

US Department of Energy [USDOE] (1990b). *First Interim Report of the Interagency Commission on Alternative Motor Fuels.* Washington, DC.

US Environmental Protection Agency [EPA] (1985). *Compilation of Air Pollutant Emission Factors, Vol. II, Mobile Sources.* Fourth edition, AP-42, Office of Mobile Sources, Research Triangle Park, NC.

US Environmental Protection Agency [EPA] (1988). *Compilation of Air Pollutant Emission Factors, Vol. II, Stationary Sources.* Fourth edition, AP-42, Office of Air and Radiation, Research Triangle Park, NC.

US National Research Council [NRC] (1992). *Automotive Fuel Economy: How Far Should We Go?* National Academy Press, Washington, DC.

Weldmann, K., H. Menrad, K. Reders and R.C. Hutcheson (1988). *Diesel Fuel Quality Effects on Exhaust Emissions.* Society of Automotive Engineers, Inc., Technical Paper Series, No. 881649, Warrendale, PA.

Wright, D. (1991). *Biomass: A New Future?* Commission of the European Communities, Forward Studies Unit, Brussels.

d'Zvrilla, D. (1991). "Reformulated Gasoline and Transportation Alternatives". Paper presented at the Conference on Energy and the Environment in the Asia-Pacific Region: Planning for an Uncertain Future, East-West Centre, Honolulu, Hawaii, 12th-15th February.

GLOSSARY

AC	alternating current
BSFC	brake specific fuel consumption
Btu	British thermal unit (1 btu = 1.055 kW)
CAFE	corporate average fuel economy
CARB	California Air Resources Board
cc	cubic centimetre
CEC	Commission of the European Community
CO$_2$	carbon dioxide
CFCs	chlorofluorocarbons
CNG	compressed natural gas
DC	direct current
DI	direct injection
EC	European Community
EGR	exhaust gas recirculation
EPA	United States Environmental Protection Agency
FTP	Federal Test Procedure
g	gram
GDP	gross domestic product
GES	Gesellshaft für Elektrische Strassenverkehr
GJ	gigajoule (one billion joules)
GWP	global warming potential
hp h	horsepower-hour (1hp h = 0.7457 kWh)
IDI	indirect injection
IEA	International Energy Agency
kg	kilogram
kJ	kilojoule

km	kilometre
ktoe	thousand metric tons of oil equivalent
kWh	kilowatt-hour (1 kWh = 3.6 MJ)
LNG	liquefied natural gas
LPG	liquefied petroleum gas
M100	fuel containing 100% methanol
M85	fuel containing 85% methanol, 15% gasoline
MBtu	million British thermal units
MJ	megajoule (one million joules)
MTBE	methyl-tertiary-butyl ether
Mtoe	million metric tons of oil equivalent
NO_x	nitrogen oxides
N_2O	nitrous oxide
OECD	Organisation for Economic Co-operation and Development
PJ	petajoule (1 PJ = 1000 TJ)
R&D	research and development
RME	rapeseed dimethyl ester
RWE	Rheinish-Westfalisches Elektrizitatswerk
SCAQMD	South Coast Air Quality Management District
TJ	terajoule (1 TJ = 1000 GJ)
toe	metric ton of oil equivalent (1 toe = 41 868 kJ)
TPES	total primary energy supply
UN-ECE	United Nations Economic Commission for Europe
VOC	volatile organic compounds
Wh	watt-hour

MAIN SALES OUTLETS OF OECD PUBLICATIONS
PRINCIPAUX POINTS DE VENTE DES PUBLICATIONS DE L'OCDE

ARGENTINA – ARGENTINE
Carlos Hirsch S.R.L.
Galería Güemes, Florida 165, 4° Piso
1333 Buenos Aires   Tel. (1) 331.1787 y 331.2391
Telefax: (1) 331.1787

AUSTRALIA – AUSTRALIE
D.A. Information Services
648 Whitehorse Road, P.O.B 163
Mitcham, Victoria 3132   Tel. (03) 873.4411
Telefax: (03) 873.5679

AUSTRIA – AUTRICHE
Gerold & Co.
Graben 31
Wien I   Tel. (0222) 533.50.14

BELGIUM – BELGIQUE
Jean De Lannoy
Avenue du Roi 202
B-1060 Bruxelles   Tel. (02) 538.51.69/538.08.41
Telefax: (02) 538.08.41

CANADA
Renouf Publishing Company Ltd.
1294 Algoma Road
Ottawa, ON K1B 3W8   Tel. (613) 741.4333
Telefax: (613) 741.5439
Stores:
61 Sparks Street
Ottawa, ON K1P 5R1   Tel. (613) 238.8985
211 Yonge Street
Toronto, ON M5B 1M4   Tel. (416) 363.3171

Les Éditions La Liberté Inc.
3020 Chemin Sainte-Foy
Sainte-Foy, PQ G1X 3V6   Tel. (418) 658.3763
Telefax: (418) 658.3763

Federal Publications
165 University Avenue
Toronto, ON M5H 3B8   Tel. (416) 581.1552
Telefax: (416) 581.1743

Les Publications Fédérales
1185 Avenue de l'Université
Montréal, PQ H3B 3A7   Tel. (514) 954.1633
Telefax : (514) 954.1633

CHINA – CHINE
China National Publications Import
Export Corporation (CNPIEC)
16 Gongti E. Road, Chaoyang District
P.O. Box 88 or 50
Beijing 100704 PR   Tel. (01) 506.6688
Telefax: (01) 506.3101

DENMARK – DANEMARK
Munksgaard Export and Subscription Service
35, Nørre Søgade, P.O. Box 2148
DK-1016 København K   Tel. (33) 12.85.70
Telefax: (33) 12.93.87

FINLAND – FINLANDE
Akateeminen Kirjakauppa
Keskuskatu 1, P.O. Box 128
00100 Helsinki   Tel. (358 0) 12141
Telefax: (358 0) 121.4441

FRANCE
OECD/OCDE
Mail Orders/Commandes par correspondance:
2, rue André-Pascal
75775 Paris Cedex 16   Tel. (33-1) 45.24.82.00
Telefax: (33-1) 45.24.81.76   or (33-1) 45.24.85.00
Telex: 640048 OCDE

OECD Bookshop/Librairie de l'OCDE :
33, rue Octave-Feuillet
75016 Paris   Tel. (33-1) 45.24.81.67
(33-1) 45.24.81.81

Documentation Française
29, quai Voltaire
75007 Paris   Tel. 40.15.70.00
Gibert Jeune (Droit-Économie)
6, place Saint-Michel
75006 Paris   Tel. 43.25.91.19
Librairie du Commerce International
10, avenue d'Iéna
75016 Paris   Tel. 40.73.34.60
Librairie Dunod
Université Paris-Dauphine
Place du Maréchal de Lattre de Tassigny
75016 Paris   Tel. 47.27.18.56
Librairie Lavoisier
11, rue Lavoisier
75008 Paris   Tel. 42.65.39.95
Librairie L.G.D.J. - Montchrestien
20, rue Soufflot
75005 Paris   Tel. 46.33.89.85
Librairie des Sciences Politiques
30, rue Saint-Guillaume
75007 Paris   Tel. 45.48.36.02
P.U.F.
49, boulevard Saint-Michel
75005 Paris   Tel. 43.25.83.40
Librairie de l'Université
12a, rue Nazareth
13100 Aix-en-Provence   Tel. (16) 42.26.18.08
Documentation Française
165, rue Garibaldi
69003 Lyon   Tel. (16) 78.63.32.23
Librairie Decitre
29, place Bellecour
69002 Lyon   Tel. (16) 72.40.54.54

GERMANY – ALLEMAGNE
OECD Publications and Information Centre
August-Bebel-Allee 6
D-W 5300 Bonn 2   Tel. (0228) 959.120
Telefax: (0228) 959.12.17

GREECE – GRÈCE
Librairie Kauffmann
Mavrokordatou 9
106 78 Athens   Tel. 322.21.60
Telefax: 363.39.67

HONG-KONG
Swindon Book Co. Ltd.
13–15 Lock Road
Kowloon, Hong Kong   Tel. 366.80.31
Telefax: 739.49.75

HUNGARY – HONGRIE
Euro Info Service
kázmér u.45
1121 Budapest   Tel. (1) 182.00.44
Telefax : (1) 182.00.44

ICELAND – ISLANDE
Mál Mog Menning
Laugavegi 18, Pósthólf 392
121 Reykjavik   Tel. 162.35.23

INDIA – INDE
Oxford Book and Stationery Co.
Scindia House
New Delhi 110001   Tel.(11) 331.5896/5308
Telefax: (11) 332.5993
17 Park Street
Calcutta 700016   Tel. 240832

INDONESIA – INDONÉSIE
Pdii-Lipi
P.O. Box 269/JKSMG/88
Jakarta 12790   Tel. 583467
Telex: 62 875

IRELAND – IRLANDE
TDC Publishers – Library Suppliers
12 North Frederick Street
Dublin 1   Tel. 74.48.35/74.96.77
Telefax: 74.84.16

ISRAEL
Electronic Publications only
Publications électroniques seulement
Sophist Systems Ltd.
71 Allenby Street
Tel-Aviv 65134   Tel. 3-29.00.21
Telefax: 3-29.92.39

ITALY – ITALIE
Libreria Commissionaria Sansoni
Via Duca di Calabria 1/1
50125 Firenze   Tel. (055) 64.54.15
Telefax: (055) 64.12.57
Via Bartolini 29
20155 Milano   Tel. (02) 36.50.83
Editrice e Libreria Herder
Piazza Montecitorio 120
00186 Roma   Tel. 679.46.28
Telefax: 678.47.51
Libreria Hoepli
Via Hoepli 5
20121 Milano   Tel. (02) 86.54.46
Telefax: (02) 805.28.86
Libreria Scientifica
Dott. Lucio de Biasio 'Aeiou'
Via Coronelli, 6
20146 Milano   Tel. (02) 48.95.45.52
Telefax: (02) 48.95.45.48

JAPAN – JAPON
OECD Publications and Information Centre
Landic Akasaka Building
2-3-4 Akasaka, Minato-ku
Tokyo 107   Tel. (81.3) 3586.2016
Telefax: (81.3) 3584.7929

KOREA – CORÉE
Kyobo Book Centre Co. Ltd.
P.O. Box 1658, Kwang Hwa Moon
Seoul   Tel. 730.78.91
Telefax: 735.00.30

MALAYSIA – MALAISIE
Co-operative Bookshop Ltd.
University of Malaya
P.O. Box 1127, Jalan Pantai Baru
59700 Kuala Lumpur
Malaysia   Tel. 756.5000/756.5425
Telefax: 757.3661

MEXICO – MEXIQUE
Revistas y Periodicos Internacionales S.A. de C.V.
Florencia 57 - 1004
Mexico, D.F. 06600   Tel. 207.81.00
Telefax : 208.39.79

NETHERLANDS – PAYS-BAS
SDU Uitgeverij
Christoffel Plantijnstraat 2
Postbus 20014
2500 EA's-Gravenhage   Tel. (070 3) 78.99.11
Voor bestellingen: Tel. (070 3) 78.98.80
Telefax: (070 3) 47.63.51

**NEW ZEALAND
NOUVELLE-ZÉLANDE**
Legislation Services
P.O. Box 12418
Thorndon, Wellington   Tel. (04) 496.5652
Telefax: (04) 496.5698

OECD PUBLICATIONS, 2 rue André-Pascal, 75775 PARIS CEDEX 16
PRINTED IN FRANCE
(61 93 02 1) ISBN 92-64-13804-8 - N° 46244 1993